PRENTICE-HALL

Foundations of Cultural Geography Series
PHILIP L. WAGNER, *Editor*

THE LOOK OF THE LAND,
John Fraser Hart

GEOGRAPHY OF DOMESTICATION,
Erich Isaac

HOUSE FORM AND CULTURE,
Amos Rapoport

GEOGRAPHY OF RELIGIONS,
David E. Sopher

ENVIRONMENTS AND PEOPLES,
Philip L. Wagner

THE CULTURAL GEOGRAPHY OF THE UNITED STATES,
Wilbur Zelinsky

A PROLOGUE TO POPULATION GEOGRAPHY,
Wilbur Zelinsky *

RICHARD E. DAHLBERG, *Series Cartographer*

* *In Prentice-Hall's Foundations of Economic Geography Series, also.*

Foundations of Cultural Geography Series

RICHARD E. DAHLBERG, *Series Cartographer*

The
Look
of the Land

JOHN FRASER HART

University of Minnesota

PRENTICE-HALL, INC., Englewood Cliffs, N.J.

Library of Congress Cataloging in Publication Data

HART, JOHN FRASER.
    The look of the land.

    (Foundations of cultural geography series)
    Includes bibliographical references and index.
    1. Rural geography.   2. United States—Description
and travel.   3. Agricultural geography—United States.
I. Title.
G127.H29   910'.09'1734   74-20995
ISBN   0-13-540534-3
ISBN   0-13-540526-2 pbk.

PRENTICE-HALL INTERNATIONAL, INC., *London*
PRENTICE-HALL OF AUSTRALIA, PTY. LTD., *Sydney*
PRENTICE-HALL OF CANADA, LTD., *Toronto*
PRENTICE-HALL OF INDIA PRIVATE LIMITED, *New Delhi*
PRENTICE-HALL OF JAPAN, INC., *Tokyo*

The ideal geographer should be able to do two things:
he should be able to read his newspaper with understanding,
and he should be able to take his country walk —
or maybe his town walk—with interest.

H. C. DARBY, *inaugural lecture,*
University of Liverpool, 1946

Foundations of Cultural Geography Series

The title of this series, Foundations of Cultural Geography, represents its purpose well. Our huge and highly variegated store of knowledge about the ways that humans occupy and use their world becomes most meaningful when studied in the light of certain basic questions. Original studies of such basic questions make up this series of books by leading scholars in the field.

The authors of the series report and evaluate current thought centered on the questions: How do widely different systems of ideas and practice influence what people do to recreate and utilize their habitats? How do such systems of thought and habitat spread and evolve? How do human efforts actually change environments, and with what effects?

These questions are approached comparatively, respecting the great range of choice and experience available to mankind. They are treated historically as well, to trace and interpret and assess what man has done at various times and places. They are studied functionally, too, and whatever controlling processes and relationships they may reveal are sought.

Diverse tastes and talents govern the authors' attack on these problems. One deals with religion as a system of ideas both influencing and reflecting environmental conditions. Another evaluates the role of belief and custom in reshaping plant and animal species to human purposes. Some consider the use and meaning of human creations, like houses or cities, in geographic context; others treat of the subtle and complex relationships with nature found in agricultural systems of many sorts. One author looks at an entire country as a culturally-shaped environment; another analyzes the mechanics of the spread of customs and beliefs in space. All work toward an understanding of the same key problems. We invite the reader to participate actively in the critical rethinking by which scholarship moves forward.

PHILIP L. WAGNER

John Fraser Hart's masterly essay on THE LOOK OF THE LAND goes as far as anything I know toward conveying the authentic feel of geographic work in the field. Long a frequenter of country crossroad stores and an explorer of old barns and barnyards, Professor Hart combines experience and judgment with a contagious enthusiasm and a keen observant eye, as he leads his reader on a new kind of trip through rural America.

PHILIP L. WAGNER

Contents

Preface

Some years ago, in response to a not unreasonable criticism that my interests in geography seemed to be riding off in all directions, I evolved a framework into which most of them seemed to fit fairly comfortably; that framework became first a single lecture, then a course, and finally this book. In it I have tried to avoid repeating too much of what I have already published elsewhere, in the assumption that anyone who is interested can follow up the citations in footnotes. I have preferred here to try to relate the ideas of others to my framework, and to make rather brief and tentative forays into some of the less well explored areas, in the hope that others might be tempted to help fill some of the gaps I perceive.

Many people have helped me, and I should like to thank each one, but I must express special appreciation to E. Estyn Evans, J. B. Jackson, Peirce Lewis, Cotton Mather, Carl O. Sauer, and Glenn T. Trewartha; to Dr. George Mann and his colleagues; to Pat Burwell and Sandy Haas; to Arlette Lindbergh; and to Meredith, who has kept reminding me.

JOHN FRASER HART

CHAPTER 1 *the plant cover*

The United States is a nation of city-dwellers. In 1870 three-quarters of our forebears lived in rural areas, but in 1970 three-quarters of us lived in cities. For many of us today the countryside is a strange, exotic, perhaps even frightening place. We seldom bother to look at it as we whiz through it, and often when we do look we don't really understand what we see, which is unfortunate, because we have interest, affection, and concern only for things we understand and appreciate. Too many people are distracted by what is strange, bizarre, and freakish, and their basic appreciations have never been developed.

The countryside warrants far more attention than most of us have given it. The only proper way to learn about and understand it is to live in it, look at it, think about it, contemplate it, speculate about it, and ask questions about it. Reading about it is a very poor substitute indeed, but I have written this book to provide some initial clues as to the kinds of things to look for and ponder. I have emphasized things you can see, and I have also emphasized the vernacular, the common, ordinary, everyday things of the people who live on the land, because these folk features make the countryside what it is.

Landforms

Most rural landscapes are so complex that they must be analyzed in terms of their three principal components: the nature of the land surface, the kind of plant life which covers it, and the structures which man has placed upon it. In most rural areas the most immediately impressive component is the nature of the land surface, whether it be barren mountain ranges shimmering in the heat of the California desert, the stark cliff and canyon lands of the Colorado Plateau, majestic jagged peaks in the Rocky Mountains, the endless flatness of the Great Plains, the rolling glacial plains of the Middle West, tangled mountain fastnesses in Appalachia, or the gentle, glacially rounded hills of New Engand.

The study of the features of the land surface has traditionally been an integral aspect of geography. Both geomorphologists and physical geographers have written excellent textbooks which are readily available to anyone who wishes to learn about the land surface, and how it came to be as it is. Despite the very great importance of the surface features of the land in most rural landscapes, they will receive only scant attention here. In part this is because they have been treated in such great detail elsewhere, but also in part it is because they may be taken as "givens," unchangeables, in any consideration of human activities. True enough, in human times great changes have been wrought in some relatively restricted areas by the ravages of volcano, earthquake, flood, meander, and even soil erosion, but for most of the world's land surface the "everlasting hills" have been been everlasting and unchanged within the memory of man on the earth.[1]

The unraveling of the history of some part of the earth's surface, from the deepest geologic past to the present, is a challenging and rewarding exercise for the specialist, but the nonspecialist, however much he enjoys and learns from the results of his colleagues' endeavors, commonly must be content to accept the nature of the land surface as a "given" as far as man and his works are concerned. The remainder of this chapter, therefore, will be devoted to a consideration of the importance of plant life as the second component of the rural landscape, and the remainder of the volume will be devoted to consideration of the third element, the structures which man has added to it.

Plant Cover

The second principal component of any rural landscape is the kind of vegetation which covers the surface of the land, whether it consists of cultivated or noncultivated plants; of grasses, herbs, shrubs, or trees; or of some mixture of these four basic categories. The vegetation of any area is significant for at least three reasons: it contributes to the appearance of the present landscape; it can tell us a great deal about the climate which environs it and the soil which nourishes it; and it may also reveal much about the manner in which man has used the area in the past, and its potential for his use in future.

There can be few parts of the earth's surface whose ensemble of plants has not been influenced to some degree by such human activities as clearance and cultivation, the introduction of exotic plant species, the abandonment of formerly cultivated land, the depredations of grazing animals, or the ravages of fire. Speculation about "natural vegetation," and whether there is or ever has been any such thing, may be an interesting

---

[1] Richard Hartshorne, *Perspective on the Nature of Geography,* Association of American Geographers, Monograph Series, No. 1 (Skokie, Ill.: Rand McNally, 1959), pp. 84–86.

academic exercise, but the student of landscape is more interested in the "real vegetation," the kinds and types of plants which cover the land right now.[2] This real vegetation reflects the activities of man, but it also reflects the interplay of a complex of physical factors, such as elevation above sea level, slope, air drainage, the form and availability of plant nutrients in the soil, the degree of soil erosion, and the nature and periodicity of such climatic variables as temperature conditions, the availability of light and water, and exposure to sunlight and prevailing winds.

The presence and abundance of each plant species express a set of environmental factors more or less unique to that species. A community of plants fairly well summarizes all of the ecologically relevant factors and interactions, plus the effects of human activities. This is what Professor Carl Sauer meant when he remarked that "plants ought to talk to the geographer." For example, the widespread distribution of *Poa pratensis*, better known as Kentucky bluegrass, can tell him the success story of a hardy immigrant which thrived in the New World once man had helped it across the Atlantic Ocean. Douglas firs west of the Cascades, western white pines in the northern Rockies, and various pine trees in the Southeast tell of the forest fires which have swept these areas.

"Pin oak flats" on claypan soils of the lower Ohio valley; "cedar glades" on thin-soiled limestone and dolomite areas in Appalachia; "old field pines" on eroded and abandoned farmlands in the eastern and southern United States; "yard plants" which still mark the settlement sites of long-vanished Caribbean Indian groups; fields of soybeans in southern Indiana on bottomland flooded too late in the spring to replant corn; even rows of cedar trees planted by bluejays along fence lines in various parts of the eastern United States; each of these has a tale to tell to the observant geographer, just as each makes its own distinctive contribution to the landscape.

Between 1910 and 1959 the area of cleared farmland in the eastern United States declined by more than forty million acres, an area considerably larger than the entire state of Iowa (Fig. 1–1). On the Appalachian fringe of Kentucky, as in other parts of the South, the best indicator of farmland abandonment is the kind of vegetation that covers a piece of ground.[3] As a general rule, every pasture in this area has to be "clipped" regularly with a mowing machine to keep down weeds and brush. Any pasture which is not clipped at least once or twice a year

[2] A. W. Küchler, *Manual to Accompany the Map: Potential Natural Vegetation of the Conterminous United States*, Special Publication No. 36 (New York: American Geographical Society, 1964), pp. 1–6. May Theilgaard Watts, *Reading the Landscape: An Adventure in Ecology* (New York: Macmillan, 1957), is required reading for anyone interested in understanding the plants which cover the land; this is one of the most popular books on the reading list for my course on "The Look of the Land," but of course there are always a few grouches who don't like anything a professor makes them read. An equally delightful book is May Theilgaard Watts, *Reading the Landscape of Europe* (New York: Harper and Row, 1971).

[3] John Fraser Hart, "Abandonment of Farm Land on the Appalachian Fringe of Kentucky," *Mélanges de Géographie, Physique, Humaine, Economique, Appliquée, Offerts à M. Omer Tulippe* (Gembloux, Belgium: Editions J. Duculot, 1967), Vol. I, pp. 352–60.

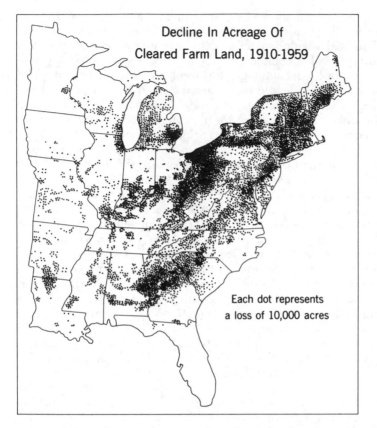

FIG. 1–1.   Decline in acreage of cleared farmland in the eastern United States, 1910–1959. Reproduced by permission from the Annals, Association of American Geographers, *Vol. 58 (1968), 425.*

will soon be invaded by broom sedge, a coarse, brown, knee-high plant. Broom sedge is followed in a year or two by blackberry briers, sumac bushes, sassafras and persimmon sprouts, and eventually by young red cedar trees. Within a decade or two the old field will be covered by a second growth of forest, and you would hardly know that the land had ever been cultivated unless you have learned to read the story told by the plants.

Clearing the Woodlands of Europe

"Perhaps the greatest single factor in the evolution of the European landscape," said H. C. Darby, "has been the clearing of the wood that once clothed almost the entire continent." [4] Darby believed that neolithic

[4] H. C. Darby, "The Clearing of the Woodland in Europe," in William L. Thomas, ed., *Man's Role in Changing the Face of the Earth* (Chicago: University of Chicago Press, 1956), pp. 183–216.

and Roman occupance of the forests of western Europe may have produced some agricultural clearings, and small scattered iron smelting operations may have consumed substantial quantities of wood, but that the real attack on these woodlands occurred between the fifth and the eleventh centuries.

England still contained sizable tracts of woodland at the time of the Domesday Book (1086), but these probably filled the gaps between villages, because most present-day villages had already been founded, and no large forested areas remained to be cleared. The medieval village was a cluster of houses and farm buildings surrounded by great stretches of open cropland which had neither fences nor hedges. Most of the open fields were not enclosed until the period between 1450 and 1850, but since enclosure trees have grown up along many hedgerows to produce what many people think of as the "typical English landscape."

Through the centuries Englishmen continued to cut down trees for building and domestic purposes; for such industrial needs as pit-props for mining, and charcoal for smelting; for the expansion of crop and pasture land; and for building the ships needed for expansion overseas. Conflicts inevitably developed between those who wished to use the forests for different purposes, and some writers had begun to express their concern about timber shortages as early as the sixteenth century. Many landowners began planting trees in the seventeenth and eighteenth centuries, but not all were economically motivated. Some planted coverts to provide shelter for foxes and other game, some planted decorative spinneys and ornamental belts to beautify the countryside, and some scattered trees through the parks which surrounded their country houses.

The development of scientific forestry on any sizable scale, however, had to await the creation of government forestry services, whose activities have been restricted, for the most part, to the poorer and less fertile lands, where conifers thrive better than hardwoods. Large stretches of infertile sand and gravel in northwestern Germany, in the Breckland of eastern England, in the Belgian Kempenland, and in the Landes of southwestern France have been converted from heather-covered heathlands into forests of pine, spruce, fir, and larch. In Britain extensive coniferous forests have also been planted on upland moorlands, whose rough grazing lands had hitherto been used almost entirely for sheep ranching.[5] British foresters have learned that existing plant communities are the best indicators of site suitability when they choose the right species of trees to plant on these treeless areas.

The distinctive vegetation of the Mediterranean lands of southern Europe provides striking evidence of man's impact on the plant cover.[6] The original forests have been reduced to a few scattered remnants, and the distribution of woodland has become almost the obverse of the dis-

[5] John Fraser Hart, *The British Moorlands: A Problem in Land Utilization* (Athens, Ga.: University of Georgia Press, 1955), pp. 47–76.
[6] Marvin W. Mikesell, *Northern Morocco: A Cultural Geography,* University of California Publications in Geography, Vol. 14 (Berkeley and Los Angeles: University of California Press, 1961), pp. 95–119; and *idem,* "The Deforestation of Mount Lebanon," *Geographical Review,* Vol. 59 (1969), 1–28.

tribution of man and his animals. The present scrub vegetation (maquis) is a degenerate form of the original oak woodland, and maquis, in turn, may be converted to a more open scrub formation (garigue) by repeated burning or overgrazing.

The expansion and contraction of scrub vegetation is closely correlated with the changing distribution and intensity of human activities. Man has changed forest to scrub by cutting wood for domestic and industrial purposes, by burning to improve pastures or to clear fields for cultivation, and by permitting overgrazing by sheep and goats. Numerous lines of evidence have to be explored in order to decipher the character and evolution of the Mediterranean vegetation. The original forests have been almost completely destroyed by cutting, burning, and grazing, and their pattern can be reconstructed by an examination of historical evidence and by careful observation of the few stands which survive, especially those on sacred sites such as cemeteries and the shrines of saints. Such sites normally are protected from animal, fire, and axemen; their plant growth is denser than on adjacent unprotected land, and the individual plants grow taller and thicker than plants of the same species on unprotected sites.

Plants as Ecological Indicators

In North America geographers have been less interested in the history of forest clearance than in the use of plants as indicators of environmental potential. The growth of individual plants and plant ensembles is so intimately related to the physical nature of the sites upon which they grow that the present vegetation is an excellent indicator, sometimes a bit general but often quite specific, of what might reasonably be expected of other plants if they were grown in the same place. At one site, for example, the vegetation may hoist a warning signal to the intending settler, but at another it may lay out the welcome mat to inform him that conditions are suitable for the plants he hopes to cultivate.

The early immigrants to North America were very well aware of this fact, and they selected their land according to the kinds of trees which grew upon it.[7] Kiefer has shown how the pioneer settlers in the eastern part of the Middle West used the vegetation to appraise the quality of the land.[8] The southeastern part of Rush County, Indiana,

---

[7] Friedrich Ratzel, *Politische und Wirtschafts-Geographie der Vereinigten Staaten von Amerika* (Muenchen: R. Oldenbourg, 1893), p. 413. Most handbooks written for European immigrants to the United States in the nineteenth century contained at least one full chapter explaining the significance of the different hardwood tree species as indicators of soil quality, and no less than 50 extensively illustrated pages are devoted to this topic in C. L. Fleischmann, *Der Nordamerikanische Landwirth: Ein Handbuch fur Ansiedler in den Vereinigten Staaten* (Frankfurt: C. F. Heyer Verlag, 1852), pp. 21–71.

[8] Wayne E. Kiefer, *Rush County, Indiana: A Study in Rural Settlement Geography*, Geographic Monograph Series, Vol. 2 (Bloomington, Ind.: Indiana University Department of Geography, 1969).

for example, is underlain by a sheet of older Wisconsin glacial till that has been subjected to considerable stream erosion and has a well-drained rolling surface, whereas the newer till in the northwest is still level and poorly drained. Differences in the natural drainage of the two till sheets were reflected by differences in vegetation which were well-known and clearly understood by the early settlers. The better drained areas, which were dominated by sugar maple and oak, were known as sugar-tree land, whereas the wet swampy ground was known as beech land. "We cannot be certain that every purchaser used the forest cover as an index of land quality, but there is good reason to believe that many of them did so, and even better reason to believe that they were justified in doing so." [9]

Early settlers in Rush County preferred the well-drained sugar-tree land, the rolling areas, and avoided the poorly drained beech land until most of the better land had been bought (Fig. 1–2). Approximately half of the county was purchased between October, 1820, when sales were started at the Land Office in Brookville, and the end of the year 1824. Apart from a Quaker settlement in the northwest, this land was concentrated in the southeastern part of the county. Most of the early settlers entered the county from the southeast, and it is hard to determine how much this fact is related to the location of early land purchases, but it is nonetheless remarkable that three-quarters of the county's sugar-tree land, and less than a third of its beech land, were purchased in the first four years after the land was opened for sale.

Most of the early settlers in Rush County found a kind of vegetation with which they were already familiar. An unfamiliar kind of vegetation, however, might create problems for the new settler, as witness the early confusion regarding the value of prairie grasslands. Although today these are some of the most productive agricultural lands in the world, it is often supposed that the early settlers of the Middle West, and especially Southerners, were strongly prejudiced against these tree-less lands; at one time, in fact, they were called "barrens." Sauer has demonstrated, however, that early settlers in Kentucky attached no odium to the term "barren," but used it because the English language does not have any suitable name for a treeless grassland. [10]

In the Middle West the name "barren" was soon replaced by "prairie," borrowed from the French, but both referred only to treeless areas, not to the quality of their soil; old-timers in the richest parts of Iowa still talk proudly of having grown up on the "barren prairie." The initial avoidance of prairie lands was primarily a reflection of the difficulties a treeless area created for people who relied on wood to build their houses, their barns, their fences, and perhaps most important of all, their fires. Furthermore, it is quite clear that the early settlers did not really try to avoid the prairies; studies in Iowa, in Illinois, and in Texas

[9] Kiefer, *Rush County, Indiana*, footnote 8, pp. 47–48.
[10] John Leighly, ed., *Land and Life: A Selection from the Writings of Carl Ortwin Sauer* (Berkeley and Los Angeles: University of California Press, 1963), pp. 23–24.

have shown that the pioneers, wherever possible, preferred to settle along the prairie-woodland edge, so that they might enjoy the advantages of both types of country.[11]

Perhaps geographers, in their preoccupation with the physical environment, might have paid too much attention to the vegetation of the prairies, and not enough to their accessibility. The vegetation clearly was important, but a pioneer was not likely to settle land he could not reach, and perhaps less desirable land near routeways was settled before better land at a distance. This idea is difficult to test, because the distribution of woodland and prairie accords so closely with the flow of internal migration: the pioneers moved from east to west, from woodland toward prairie, and their principal routeways were the rivers whose valleys carried the westernmost tentacles of woodland. The early settlers certainly used the wooded valleys to bypass upland prairies, but I know of no area where they leap-frogged directly across a prairie to reach wooded areas on the other side, and I do know of at least one area, in southeastern Minnesota, where upland prairies were settled before wooded valleys nearby because a road cut across the prairie.[12]

Cultivated Plants and Crop Rotations

The history of woodland clearance is interesting, and plants can be useful ecological indicators, but the principal contribution of plant life to the look of the land, as far as most of us are concerned, is made by the plants which man has cultivated. The cultivated plants, in the broadest sense, include the trees, shrubs, flowers, and grass with which we adorn the lots upon which our houses stand, as well as the crops of the fields, many of the grasses of pasture and range, and many of the trees of orchard and woodland. The plants of yard and garden are much too complex for treatment here, however, and we will stick to field crops.

Field crops are important to all of us, because animals, including the human animal, cannot eat rocks. Plants are vital intermediaries between the mineral and animal worlds. They convert the minerals of the soil into food which animals can digest. The forms and numbers of animal life which can inhabit any area are determined, in large measure, by the assemblage of plants which can be supported by the climate and

---

[11] Leslie Hewes, "Some Features of Early Woodland and Prairie Settlement in a Central Iowa County," *Annals,* Association of American Geographers, Vol. 40 (1950), 40–57; B. P. Birch, *The Environment and Settlement of the Prairie-Woodland Transition Belt: A Case Study of Edwards County, Illinois,* Southampton Research Series in Geography, 6 (Southampton, England: Southampton University Department of Geography, 1971); and Terry G. Jordan, "Pioneer Evaluation of Vegetation in Frontier Texas," *Southwestern Historical Quarterly,* Vol. 76 (1973), 233–54.

[12] Charles E. Dingman, "Land Alienation in Houston County, Minnesota: Preferences in Land Selection," *The Geographical Bulletin,* Vol. 4 (1972), 45–49.

soil of the area. Perhaps a few folk still live mainly on what they can gather from wild plants, but most of the world's people derive their sustenance from cultivated crops, whether eaten directly or fed to animals for conversion into meat and milk.

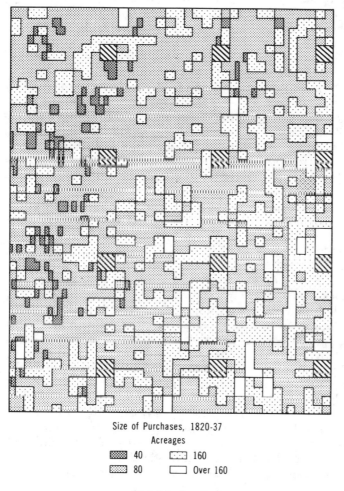

Size of Purchases, 1820-37

Acreages

▨	40	▥	160
▨	80	☐	Over 160

◪ School section

FIG. 1–2. *Size of tracts purchased in Rush County, Indiana, 1820–1937. The poorly drained beech land in the northwestern part of the county was purchased later and in smaller tracts than the rolling sugar-tree land in the southeast. Redrafted, with the author's permission, from Wayne E. Kiefer, "The Rural Settlement Geography of Rush County, Indiana," unpublished doctoral dissertation, Indiana University, 1967, p. 68.*

Geographers have made many maps showing the distribution of individual crops, but they have not paid enough attention to patterns of crop rotation, which provide the key to an understanding of the distribution of many cultivated crops in the middle latitudes.[13] Crops are grown in regular succession on the same piece of ground for numerous reasons. Some crops simply do not yield well after one particular crop, but are quite happy after another; others need to follow a "nurse crop" which protects them when they are young. Most farmers do not like to "put all of their eggs in one basket"; the farmer who grows several crops can spread the risk of loss because of bad weather or low prices, and he can also spread his workload more evenly through the year. Growing the same crop year after year in the same field can exhaust the fertility of its soil and increase the possibility of damage by diseases and insects.[14] Weeds are easy to control in some crops, but almost impossible in others. Some crops reduce the risk of soil erosion, but others increase it.

Cultivated crops fall into three basic categories from the point of view of crop rotation: row crops, hay crops, and small grain crops.

Row crops (such as corn, cotton, fodder roots, potatoes, sugar beets, soybeans, tobacco, and vegetables) are planted in rows, as the name implies. The bare ground between the rows can be intertilled or cultivated with a plow or hoe to get rid of weeds, but it is susceptible to erosion on sloping ground. Most of the row crops are cash crops, and a row crop commonly is the first, most important, and most profitable crop around which the rest of the rotation is organized.

Hay, grass, meadow, or sod crops (such as alfalfa or clover) are grazed by livestock or cut for hay. They provide the least income in many rotations, but they help to conserve soil moisture and control erosion by reducing run-off of rain water ("hay crops in the rotation are like brakes on a car; the steeper the slope, the more you need"), and leguminous hay crops can enrich the soil by adding humus and fixing nitrogen from the atmosphere. Good hay crops are hard to establish, and they need the protection of a small grain as a "nurse crop" during their early growth stages.

Small grain crops (wheat, rye, barley, and oats) are useful nurse crops for young hay plants, but they are almost equally effective in sheltering weeds. The small grains are among the oldest cultivated crops, and they were the basis for the "three-field" agricultural system which persisted through the Middle Ages in northern and western Europe.

The medieval three-field (food-feed-fallow) rotation was based on a food grain (wheat or rye) in the first year, a feed grain (barley or

13 For a discussion of crop rotations in the Corn Belt, and a listing of sources on which this discussion is based, see John Fraser Hart, "The Middle West," *Annals, Association of American Geographers*, Vol. 62 (1972), 266–68. One of the best descriptions of the medieval farming system is C. S. and C. S. Orwin, *The Open Fields*, 3rd ed. (Oxford: Clarendon Press, 1967), especially pp. 53–62.

14 Won't you sleep better tonight after being assured by M. W. Allen, W. H. Hart, and Ken Baghott that "Crop Rotation Controls Barley Root-knot Nematode at Tulelake," *California Agriculture*, June 1970, pp. 4–5?

oats) in the second, and a period of bare fallow in the third year so that the land could be plowed to get rid of its weeds. The food (or bread) grains, wheat and rye, needed a long growing season to give good harvests, and they were planted in the fall on the field which had lain fallow. The stubble of the previous year's wheat (or rye) field was plowed when winter weather conditions permitted, and in spring it was planted to a feed grain, barley or oats, which ripened for August harvest. The barley (or oats) stubble was grazed by livestock over winter, plowed in early summer, and lay fallow until planted in the fall with the start of a new rotational cycle. Wheat, the preferred bread grain, and barley, the better feed, were grown on the productive loam soils of the loesslands, but peasants on the poorer sandy lands had to make do with rye bread on their tables and oats for their livestock. The basic rotations were quite simple: wheat-barley-fallow on loam soils, rye-oats-fallow on sandy soils.

The Agricultural Revolution occurred when the medieval three-field system was modified by the introduction of row crops, such as turnips, and leguminous hay crops, such as clover. The necessity of a fallow year was eliminated by the row crops, which could be cultivated to eradicate weeds, and the legumes, which enriched the soil. The acreage of land under crops was increased by 50 percent when the third, or fallow, field was brought into production each year, and the new crops, which produced more and better winter feed, boosted agricultural production still further. A better supply of winter feed enabled farmers to keep more livestock and set in motion an upward spiral: more feed, more livestock; more livestock, more manure; more manure, richer soils; richer soils, still more feed. Although turnips were important in the early years, they are a fodder crop of relatively low value, and on many farms they have been replaced by cash crops such as sugar beets or potatoes. In contemporary Europe the standard rotation is wheat-sugar beets-barley-clover on loam soils and rye-oats-potatoes-clover on sandy soils.

German settlers brought the basic European four-year rotation to southeastern Pennsylvania, where it was modified to include corn, a crop which combined the advantages of turnips (row crop) and barley (feed for livestock). When it reached the Middle West the Pennsylvania rotation (corn-oats-wheat-clover) was changed to a three-year system of corn-small grains-hay, which is still the basic rotation of the Corn Belt. The usual small grain was winter wheat in the south, but oats in northerly areas where winter came so early that a farmer could not plant winter wheat after he had harvested his corn. As in Europe, the basic rotation had many variations. To take two simple examples, a farmer on fertile level land might take two crops of corn before planting oats (corn-corn-oats-hay), but the man on steep land subject to erosion might leave the hay crop for a second year (corn-oats-hay-hay) or even longer.

Around the margins of the Corn Belt the basic rotation was modified even more in response to increasing ecologic stress. In the hills to the south and east, for example, farmers used the standard rotation on level areas, but much of the land was too steep for any use but pasture or woodland. The plains to the west were so dry that corn and hay became risky, and all that remained of the rotation was the small grain, wheat; even wheat becomes risky if it is grown every year in dry-farming areas, and it is rotated in a two-year cycle with summer fallow. Northward the growing season was too short and too cool for corn to ripen dependably into grain before it was caught by frost, and the entire plant was cut green for silage; good soil management on the short but steep slopes of recently glaciated areas compelled many farmers to leave their land in hay for several years.

Modifications of the basic Corn Belt rotation have also been permitted or encouraged by recent changes in farm technology. Bigger and better machinery has relaxed constraints on soybeans, which used to compete with corn for labor, and on some farms soybeans have replaced small grains and hay in a new two-year cash crop rotation of corn and soybeans. Increased knowledge and use of such agricultural chemicals as nitrogenous fertilizers, weedkillers, and pesticides have reduced the necessity for crop rotation, and some farmers on good level land have begun to grow continuous corn, year after year with no intervening crops. Much as the idea may affront the purists, however, even continuous corn may be considered a form of crop rotation, although it is a far cry from the food, feed, and fallow of the Middle Ages.

 CHAPTER 2 *some basic concepts*

The third major component of the rural landscape, and the one with which we are primarily concerned in this volume, is the structures which have been added to it by man. Although these structures could be classified in many different ways, the scheme which will be used here is a simple fourfold grouping into: (1) systems of land division, land tenure, and associated features, (2) structures and ensembles of structures which are functionally related to the economy, (3) house types, and (4) the agglomeration of houses into hamlets, villages, towns, and cities. Before we begin to look at any of these groups of structures in detail, however, it will be useful to consider a few fundamental concepts which have general applicability to them all.

Morphology and Function

MORPHOLOGY.   As far as the rural landscape is concerned, the most important aspect of any man-made structure is its visible form or appearance, or its morphology. This is not to say that the student of the rural landscape must restrict himself to those things which can be seen upon it, much less to say that the purpose of geography is to study the morphology of landscape (catchy though that phrase may be!). Far from it! Unfortunately, however, some geographers have managed to work themselves into quite a lather over this issue, apparently without stopping to realize how patently ridiculous it is to become so obsessed by the form of visible objects that invisible objects are ignored, or to concentrate on morphology to the exclusion of function and process; and vice versa.

The debate which has raged over the importance of the visible in geography really seems rather pointless, because anyone who is attempting to understand a rural landscape (or any other phenomenon worthy of geographic study) must learn as much as possible about how it works, and how it came to be the way it is. It is equally true that he cannot ignore its appearance, whether on the ground or on a map, and even to suggest that he should do so is ludicrous. He obviously must begin with the way things look, their appearance or morphology, but he has failed in his mission if he goes no farther; and he has wasted his time and energy if he even bothers to argue that either approach should be followed exclusively.

FUNCTION.   Every man-made structure, even castles in the sand or smoke rings in the air, is made to serve some purpose, if only the pleasure of him who made it. We are concerned here, of course, with structures considerably less evanescent than sand castles or smoke rings, but nevertheless the student of the rural landscape, when he begins to analyze its man-made structures, must understand what the builder had in mind, what function the structure was designed to perform.

In some cases function may virtually dictate morphology. A barn built to "flue-cure" tobacco, for example, must be insulated well enough to retain the heat used in the curing process, it must have a source of heat, and it must have some means (normally flues of sheet-metal pipe) of getting the heat into the barn.[1] The barns are small and cubical, rarely more than 20 feet on a side, because the nature of the curing process requires that they be filled in a single day.

The farmer who requires additional curing capacity builds more barns rather than larger ones. Most flue-cured tobacco barns have lean-to shelters on at least one side, and about the only options the farmer has when he builds his barn are the choice of building material, color, and the number of sides on which to construct shelters.

In contrast, there are other types of structures whose morphology is not dictated by their function, whose builders may allow their fancies to roam fairly freely. What, for example, is a fence? The farmer who needs to keep his livestock in the pastures where they belong, and out of the fields where they do not, may settle for barbed wire if he has horses or cattle, but must use woven wire if he has sheep or hogs.[2] Or perhaps he may prefer a zigzag rail fence, or boards painted white, or a wall of stone, or a hedgerow, or even the roots of stumps upturned in clearing his land; he may choose from an enormous variety of structures, any one of which will satisfactorily perform the simple function of enclosing his land and his livestock.

RELICT STRUCTURES.   Sometimes the function for which a structure was built becomes unnecessary or is superseded, and then the structure

---

[1] Flue-cured tobacco barns are discussed and illustrated in John Fraser Hart and Eugene Cotton Mather, "The Character of Tobacco Barns and Their Role in the Tobacco Economy of the United States," *Annals,* Association of American Geographers, Vol. 51 (1961), 288–93.

[2] Eugene Cotton Mather and John Fraser Hart "Fences and Farms," *The Geographical Review,* Vol. 44 (1954), 202.

stands relict upon the landscape, a memento of times gone by. In the Middle West, for example, farmers who have switched from corn-hog farming to cash-grain farming no longer keep any livestock, which would require enclosure, and thus many of them have permitted their fences to fall into sad disrepair, or have even removed them completely.[3] An even more striking example of relict structures is the huge horse barns which were a necessity on virtually every farm in the Middle West before the days of the internal combustion engine; today they stand gaunt and derelict, or have been converted to other uses.

SEQUENT OCCUPANCE. Conversion to other uses, of course, is the fate of many relict structures, and this kind of change in function is often referred to as sequent occupance. The enlargement of a farm, for example, or its conversion to a different economic organization, might reduce the number of workers needed on it, and thus make available for other uses the houses in which these workers formerly lived. On an enlarging cash-grain farm in Illinois the former residence, if structurally sound, might be converted into a crib for storing the increased corn crop by nailing boards across the windows; a former sharecropper cabin might be used for hay storage on a one-time Georgia cotton plantation which had shifted to a pasture and livestock economy. It works both ways; the home and garden sections of metropolitan newspapers often carry stories of old farm buildings converted to residential use, complete with photographs of wealthy matrons lowing contentedly in the erstwhile quarters of lesser kine.

Style

Within the constraints set by the function which it must perform, the man who builds a structure may design it more or less as he sees fit. In most cases, however, the design which he selects is strongly influenced, if not actually constrained or determined, by factors such as the date of construction, the building materials which are available to him, the cultural baggage which he carries with him, and the technical competence which he possesses or can command. These factors, in combination, give the structure its style.

DATE OF CONSTRUCTION. The time at which a building was constructed places its own indelible mark, sometimes blatantly obvious, sometimes a mere recondite technical detail, upon each structure, and this mark is part of the structure's contribution to the landscape. As an illustration, consider changing fads in house types in the United States. The stately elegance of Greek Revival architecture in the first half of the nineteenth century was replaced by the gaudy ostentation of Victorian gingerbread in the latter half. This, in turn, gave way to the story-and-a-half bungalow of the Twenties, which has now been superseded by the angular ranch house.

[3] John Fraser Hart, "Field Patterns in Indiana," *The Geographical Review*, Vol. 58 (1968), 467–70.

BUILDING MATERIALS.   The stuff man has to work with plays a major role in the appearance of his structures, and much of this stuff comes directly from the natural environment of the area within which he lives. This is especially true of simpler and more primitive societies. The more advanced and sophisticated societies, operating at a higher economic level, can afford to import their building materials from considerable distances, especially for their public and ornamental buildings, but even in advanced societies the folk architecture of ordinary people tends to rely heavily on the local environment for its building materials.

The most fundamental difference between the villages in one region of England and another is their building materials. . . . Before the railways came to transport cheap slates from Wales or dump down wagon-loads of bricks in country that possessed good stone quarries, the village builder was obliged to use the natural materials which his own district afforded.[4]

The ordinary kinds of folk houses along a north-south traverse across European Russia at the turn of the century provided a classic illustration of interaction between the natural environment and the style and appearance of man-made structures.[5] The only human dwellings in the treeless northern tundra zone were rude huts, but wooden houses were common in the great northern forest. On the stone-free grassy steppes south of the forest the houses were built of earth or mud and covered with turf or thatch, whereas stone houses appeared on the stony flanks of the Caucasus Mountains, and wattle became a common building material in the warm, damp climate of the marshy Black Sea littoral.

The humble fence provides an even more prosaic illustration of the relationship between the natural environment and the materials man uses in his structures. Many fields and pastures in the lowlands of England and Ireland, for example, are stitched together by hedgerows of quickthorn, but the upland areas, where quickthorn does not flourish, have massive walls of stone, which faithfully reflect the geology of the rocks which lie beneath them (Fig. 2–1).[6] In recent years, as labor has become dearer, both stone walls and hedgerows have been patched or replaced by barbed wire, which has also been used for the majority of recent enclosures.

In eastern North America the pioneer settler constructed his first rude fence from the debris, trees, brush, and stones which he had to clear from his land before he could cultivate it, but later he erected a more permanent structure from whatever material lay closest to hand. The zigzag, snake, worm, or Virginia split-rail fence was most common

---

[4] F. R. Banks, *English Villages* (London: Batsford, 1963), p. 14.

[5] Jean Brunhes, *Human Geography*, abridged edition by Mme. M. Jean-Brunhes Delamarre and Pierre Deffontaines, trans. by Ernest F. Row (Skokie, Ill.: Rand McNally, 1952), p. 49. See also the superbly illustrated discussion in Richard Weiss, *Häuser und Landschaften der Schweiz* (Erlenbach-Zürich: Eugen Rentsch Verlag, 1969), pp. 33–61.

[6] W. R. Mead, "The Study of Field Boundaries," *Geographische Zeitschrift*, Vol. 54 (1966), 101–17.

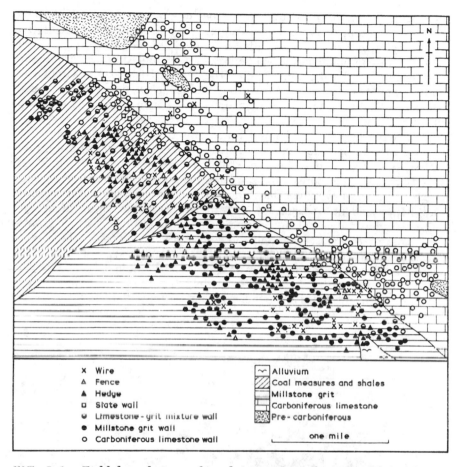

FIG. 2–1. *Field boundaries and geology in the village of Ingleton (West Riding of Yorkshire, England), based on a survey by Cedric March. Reproduced by permission of the author and publisher from W. R. Mead, "The Study of Field Boundaries,"* Geographische Zeitschrift, *Vol. 54 (1966), 114.*

in the wooded East, but stone walls enclosed the glaciated fields of New England, and uprooted stumps of white pine were turned on edge on the sandy outwash plains of Michigan and Wisconsin. Expansion westward onto the treeless prairies created some acute fencing problems for settlers from a woodland culture, and they experimented with walls of sod or mud, and a variety of hedging plants, until their miseries were put to an end by the invention of barbed wire in the 1870s.

CULTURE GROUPS. Culture, as the term is used here, refers to the learned and socially transmitted attitudes, aspirations, patterns of behavior, and technical competencies of a group of people. We are primarily interested in the values held by the culture group, and in their impact upon the landscape; some values are held more tenaciously than others, and some have greater landscape impact than others. The members of the group may be bound together by a common language, ancestry, religion, skin color, or other common ties and interests, whether alone or in combination. Snowmobilers, skiiers, and suitcase farmers are just as much culture groups as are the Scotch-Irish or the Old Order Amish, and perhaps no culture group value has had a greater impact upon the American landscape than the lawn worship which is practiced in middle-class suburbs.

Geographers in the United States, on the whole, have tended to overlook the importance, nay even the very existence, of culture groups unless those groups have been so strikingly obvious that they could not be ignored. Furthermore, some so-called fieldwork in cultural geography has been appallingly naive: the student identifies areas settled by a group, and happily identifies signs of its "influence" within those areas, but blithely ignores identical signs in adjacent areas settled by other groups.

One of the most strikingly obvious culture groups in the United States is the Amish, some 40,000 strong, who live mainly in Pennsylvania, Ohio, Indiana, and Iowa. Amish areas can be identified quite easily in plat books by shading in the areas owned by persons having no more than a dozen surnames and their variants (Figs. 2–2 and 2–3). The

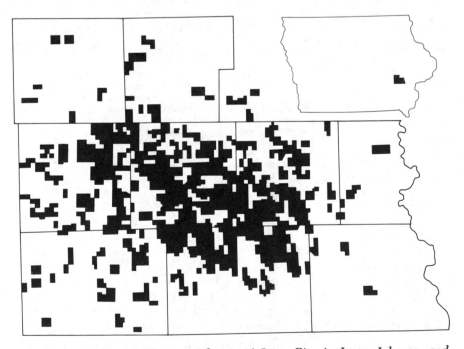

FIG. 2–2.   *The Amish area southwest of Iowa City in Iowa, Johnson, and Washington counties, Iowa. Shaded areas are owned by persons named Miller, Yoder, Bontrager, Hostetler, Gingerich, Mast, Troyer, Swartzendruber, Schlabaugh, and Beachy; Miller and Yoder alone account for 540 (62 percent) of the 868 properties. Compiled from county plat books.*

FIG. 2–3. The Amish area southeast of Goshen in Elkhart and Lagrange counties, Indiana. Shaded areas are owned by persons named Miller, Yoder, Bontrager, Hostetler, Gingerich, Mast, Troyer, Schrock, Lambright, and Eash; Miller and Yoder alone account for 472 (51 percent) of the 920 properties. Compiled from county plat books.

Amish people still cling to traditional styles of dress, and farm in the old-fashioned way, with horses rather than tractors. Road signs in Amish areas warn unwary motorists that many of the people still travel by horse and buggy, a fact amply attested to by the accumulations beside the hitching rail at the courthouse square in the county seat.

Farmsteads with two houses rather than one are a distinctive feature of Amish colonies, because the men retire at a relatively young age ("worn out by the hard labor of farming with horses," say their cynical gentile neighbors). The aging parents turn over their farm to the younger generation, and move into the "grandpa house," which may be either an addition to the main house or a completely separate structure.

Most Amish farmsteads are distinguished by Pennsylvania Dutch barns. On one side is a barn bank, an inclined driveway which leads up to a central threshing floor on the second level. The second level projects on the opposite side of the barn as a forebay, or overhang, which is cantilevered out three to six feet over the stockyard. Many of the Amish barns in Iowa lack forebays, but the necessity of a forebay on a barn is so deeply ingrained in the thinking of the Amish farmer that each barn has, in lieu of a forebay, a pent roof or overshoot which runs the length of the barn and projects out over the stockyard just as a regular forebay would have done.

The impact of the attitudes, aspirations, and patterns of behavior of a culture group (often referred to as the "cultural baggage" of the group) upon the appearance of a man-made structure might best be summed up by the statement, "That's what this group of people feel such a structure ought to look like." The importance of a group's cultural baggage, like the influence of the natural environment, commonly shows up in the folk architecture of ordinary people.[7] It may also show up in a group preference for a given crop, or a given breed of livestock.[8]

TECHNICAL COMPETENCE. The technical competence of the members of a group, or the competencies upon which they may draw, can play an important role in determining the appearance of their structures. The northern part of Milwaukee County, Wisconsin, for example, has a curious isolated cluster of octagonal barns. The late Loyal Durand, who wondered why they were there, discovered that each had been constructed under the direction of a single master barn builder who had wandered into the community and discovered a market for his talents.

Another example is provided by the shape of the roofs of barns, one of their most immediately obvious features. It has been hypothesized that roof shape might be a useful diagnostic trait of the different cultural groups which had built the barns, but Kiefer discovered that the shape simply reflected the date of construction, and increasing technical competency through time, rather than the culture of any particular group of people.[9] The straight symmetrical roof, which is easiest to construct but also the least efficient, was replaced by the gambrel roof as soon as barn builders learned how to construct it, and the gambrel roof, in turn, was replaced by the even more efficient fornicate roof.

Spatial Distribution and Areal Association

Spatial distribution and areal association are basic and fundamental concepts in all of geography, and are not restricted to an understanding of the rural landscape, but they are so relevant to such an understanding that they must be mentioned briefly. The first principal of geography is the fact that most of the things which make any place on the earth's surface the way it is—the sky, the earth, the plants, the people, their way of life, their buildings, or any aspects of these things—are distributed very unevenly over that surface; this is what we mean by *spatial distribution*. Austria, for example, has more rainfall than Australia; Bolivia has more mountains than Belgium; Ceylon has more people than Chile; Zambia has more miners than Zanzibar; and so it goes, the list is endless. The importance of spatial distribution in geography is the reason why geographers emphasize the use and importance of maps.

[7] The influence of a group's cultural baggage may be exceedingly subtle. Joseph P. Sullivan concluded that the principal feature which distinguished Irishmen's farmsteads in southern Minnesota from those of their Bohemian neighbors was use of the color green, whether on the roofs of houses and barns, or as trim around the windows and doors of the house. Any self-respecting Bohemian who purchased a farm from an Irishman was quick to change its green motif to a blue one.

[8] Karl B. Raitz and Cotton Mather, "Norwegians and Tobacco in Western Wisconsin," *Annals*, Association of American Geographers, Vol. 61 (1971), 684–96; the German tobacco-producing district north of Dayton, Ohio, which is strongly reminiscent of Lancaster County, Pennsylvania, certainly deserves more attention than it seems to have received from geographers, and where are our studies of the geography of breeds of livestock?

[9] Wayne E. Kiefer, "An Agricultural Settlement Complex in Indiana," *Annals*, Association of American Geographers, Vol. 62 (1972), 494–96.

But spatial distribution alone would not provide the basis for an intellectual discipline such as geography. If geographical variations in the things which make places the way they are had nothing in common but location, geography could be no more than an organized catalog, an encyclopedia of facts about places. But this is not the case, because the spatial distributions of most things are influenced by the spatial distributions of other things, or to put it another way, one set of things varies geographically with the geographical variations of other sets; this is what we mean by *areal association,* that many things on the surface of the earth covary geographically, and so they tend to be associated in the areas where they exist. These geographical covariations are not rigidly fixed and uniform; the combinations of things found on one part of the earth's surface differ from the combinations found in other parts, and even their interrelationships with each other differ from one part of the earth's surface to another.

The emphasis of geography, therefore, is upon the interrelatedness and interconnections of things, not upon the things themselves, nor upon their separate distribution over the face of the earth. The principal objective of geography is to study the extent and degree to which the geographical variations of any set of the things of man and nature, either in its interrelations with other sets in the same place, or in its interconnections with sets in other places, helps to determine the geographical variations of those other sets. Geographical explanations may consist of relatively simple integrations of such sets, or more complex but still partial integrations, but all attempt to approach the total integration of all interrelated phenomena. It is this total integration, the totality of areal variation, which gives particular shape to the varying character of the earth as the home of man.[10]

The sets of things selected for study in geography are usually restricted to those which help to account for the variable character of the earth's surface because their variations are interrelated with the variations of other sets, whether at the same place or in other places. This is the reason why geographers emphasize maps which show surface features, climate, plant cover, soil, land use, types of livelihood, transportation facilities, and population distribution, and why they rarely use maps of geomagnetism, fossils, gravitational anomalies, or most stratigraphic features, despite the importance of such maps for scholars in other scientific fields.

The Importance of Making a Living

Economic activities are the principal interest of many, if not most, geographers, because making a living is the most important human activity, it consumes the greatest proportion of man's time and energy, it

---

[10] Richard Hartshorne, *Perspective on the Nature of Geography,* Association of American Geographers, Monograph Series, No. 1 (Skokie, Ill.: Rand McNally, 1959), p. 106.

produces the greatest number of changes, and it probably is more closely related to other geographical variables (including population density) than are any of his other activities. Economic activities can be so important, in fact, that certain economists and economic geographers have tended to dismiss as "irrational" any activity which is not obviously motivated by economic considerations. This is the sheerest nonsense, of course, because many human activities, even in economically advanced societies, are completely rational even though they are nowise economically motivated. One of our principal interests in this volume, in fact, is "non-economic" activities, but despite our primary concern with the manner in which different groups of men with different cultures have molded and modified the rural landscape, we must accept the fact that the necessity of making a living has been the principal (but certainly not the only) agent which has produced it.

As a general rule, men cannot live where they cannot grow food. Many exceptions instantly come to mind, of course: fur trading posts, sawmills, and pulp and paper mills in the vast wilderness of the boreal forest; small fishing ports nestled in snug harbors along stern and rockbound coasts; and mining camps in bleak, forbidding areas where food, water, workers, fuel, even building materials have been brought in from outside to develop rich mineral deposits. But some of these settlements are temporary, depending upon a resource which is not renewable, and they will be abandoned when the resource is exhausted. And all of them are as but a drop in the bucket when the total population of the earth is considered, for their significance is small indeed.

A much more significant exception might be sought in the millions of city people who do not grow their own food, yet even this exception is more illusory than real, because no city could long survive without a reliable source of food, and for the majority, a source which is close at hand. The first cities of the Middle East, for example, and the great cities of Western Europe, grew up in the breadbaskets of their respective areas, and even today no sizable city lies beyond the outer limits of modern commercial agriculture. Furthermore, we must not forget that by far the greater part of the occupied surface of the earth is still rural; in 1970, for example, three-quarters of the people of the United States lived in urban areas, yet these areas covered less than 2 percent of the total land surface of the nation.

Making a living by producing food (and fiber), therefore, not only must be considered one of the most fundamental human economic activities, but it also plays a major role in shaping the rural landscape. The systems of agriculture and/or ranching in a given area are an important factor in determining the number of people which that area can comfortably support, and they probably have some influence on the way in which the people are distributed within it. An agricultural system also influences the rural landscape by its effect upon the land use and plant cover of the area, and by the kinds of man-made structures which it requires.

The Occupance of Area

How do groups of men go about occupying a rural area, and making a living there by producing food and fiber? The attempt to answer this question was responsible, in large measure, for the decision to group the man-made structures of the rural landscape into four basic categories.

SYSTEMS OF LAND DIVISION AND LAND TENURE.   At first glance it might seem strange to include systems of land division and land tenure under man-made structures, yet the essential initial step in the occupance of area is to divide up the land and to establish rights of ownership to it. This division may take place before the land is occupied, as happened in much of the United States, or it may be coincident with or even subsequent to settlement. In many areas the system of land division, as originally devised and perhaps later modified, is reflected in road patterns, and also in the pattern of political division, whether one deals with political units as large as a state or a nation, as small as a farm, or even as small as the plot of land occupied by a single family residence.

Questions of efficient utilization and tenure arise once the land has been divided and ownership rights have been established. The land must be subdivided into units which are manageable for the use to which it will be put, and this produces distinctive field patterns in the countryside and lot patterns in urban areas. The person (or group, in some cases) who acquire rights of ownership to a piece of land has at least three options in managing it: he may lease it in whole or in part to another person, he may manage all of it himself, or he may enlarge the area he manages by renting additional land from another land owner.

STRUCTURES FUNCTIONALLY RELATED TO THE ECONOMY.   Once the land has been divided up and rights of ownership and management have been established, decisions must be made by the manager or operator as to how the land will be used. These decisions will be reflected in the structures which are required if that use is to be effective. Most farmers, for example, need fences of some sort, either to enclose their livestock or to protect their crops. Livestock need shelters and handling areas such as barns, corrals, and feedlots. Most feed crops need protected storage in structures such as barn lofts, silos, corncribs, and granaries. Crops such as tobacco, hops, peas, and mint require processing in special structures on the farm. Tractors, other machinery, and tools must be protected against the elements.

Each operating unit of the agricultural economy requires a distinctive ensemble of structures such as these, which is a rather high-flown way of saying that every farm has a farmstead. Every farmstead says a great deal about the activities which are carried on by the farm operator. The observant traveler, once he understands the functions of the various farm structures, can learn much about the economy of any rural area simply by keeping a sharp eye on the complex of structures which comprise its farmsteads.

HOUSE TYPES.   The man who occupies a piece of land needs shelter for himself and his family, in addition to the structures which are required by his farm operation. The third group of man-made structures on the rural landscape, therefore, and a very important group indeed, consists of the houses which men have chosen to build for themselves. House types are placed in a separate category from other farm structures, however, for two reasons. One is the fact that house types are not nearly so closely tied to the farm operation, and any realistic examination of house types must consider those of the city as well as those of the country. The second is the fact that a house, to a far greater degree than any other man-made structure, reflects the whims and fancies of the man who built it rather than the function for which it was built. Despite some outstanding pioneering work in Europe, North America, and other parts of the world, the geographic study of house types is one of the most difficult, but mandatory, research areas in contemporary cultural geography.

RURAL SERVICE CENTERS.   All commercial rural economic systems, and many which are only marginally commercial, require some central place where goods may be collected, processed, and distributed, and where services of various sorts may be provided for the rural populace. Although the literature of urban geography is rich, in North America perhaps too little attention has been paid to the nature and function of the smaller central places which are essentially service centers for rural areas. In Europe, where many small and not-so-small clusters of houses are agricultural villages rather than rural service centers, the primary concern of geographers appears to have been the morphology of villages rather than their function, but neither should be ignored.

CHAPTER 3   *land division*
in Britain

The arrangement of man-made structures on the landscape cannot be understood without some knowledge of the history of settlement, the division of the land, and the acquisition of title to it by the settlers. One of the necessary first steps in the human occupance of any part of the earth's surface is to decide who has what rights to use which parts of it. The way in which the land is divided has a strong influence on administrative boundaries, property boundaries, field boundaries, methods of cultivation, the arrangement of farmsteads, and the layout of roads and other transportation lines. It also has some effect on systems of taxation, on the amount of litigation with which the people are afflicted, and even on their personal comfort and their attitudes toward the environment.

A concern about the division of land, and rights to use it, appears to have developed in association with the development of agriculture. Primitive hunting, fishing, and collecting groups appear to have rather vague concepts of territoriality and land division. They know the hunting grounds and collecting areas which they and their forebears have traditionally been accustomed to use, and they may even be moved to violence in defense of these areas against intruders, but otherwise their notions about rights to land are rudimentary. The agriculturalist, however, feels that he has a right to the products of the land after he has tilled the soil, planted the seed, and tended the crop. He demands that his fellows respect his right to his plot of ground and what he can produce on it, but in return he is prepared to accord equal respect to their rights to their plots.

The right to use land may be held by an individual, or it may be held in common by all members of a tribe, community, or society. This right may be exercised by the owner, or it may be extended to another user, usually in return for some consideration such as the payment of rent, whether in cash or in kind, or the performance of specified services. In order to minimize conflict between the owner and his neighbors, it is customary to demarcate the limits of any right to land by drawing a boundary around the land to which the right pertains. The overall system (or lack thereof) whereby such boundaries are drawn is a system of land division. Systems of land division fall into two broad categories: (1) those established before the land was occupied and settled have *antecedent* boundaries with regular geometric patterns; and (2) those established *coincident* with, or even *subsequent* to, settlement have irregular boundaries which follow easily distinguishable natural features such as streams and ridge lines.

Early Land Division in Britain

The British Isles provide an excellent location for examining the complexities of systems of land division established coincident with or subsequent to settlement. Not only does Britain have examples of virtually every important feature of land division which can be found in Western Europe, but it also has a distinctive Atlantic Fringe which most parts of the continent cannot match. Furthermore, most of the early settlers of North America came from the British Isles, and part of the heritage they brought with them was British ideas as to how land had been and could be divided.

Scholars have waged fierce debates, often tinged (or worse) with nationalistic overtones, about the origin, evolution, and significance of various systems of land division and land tenure in Western Europe. In a sense such debates have been justified, because every village and perhaps even every field is unique in an area with such a long and turbulent history; in Belgium, which is about the size of Delaware and Maryland combined, Dussart has been able to identify no fewer than 18 distinctive regions of field patterns.[1] There are occasions, however, when one must attempt to see the forest and ignore the trees (not to mention the branches, or even the twigs, which have so fascinated some scholars). The following discussion has not been written for the puristic specialist, nor is it an attempt to enter a debate which is apparently endless; it is merely a sketch, with broad strokes, of some of the systems of land division and land tenure which are still very much visible in the contemporary landscape of rural Britain.

The earliest recognized human inhabitants of Britain were paleolithic bands who roamed the edge of the forest in small kin groups, the men hunting wild game, and the women and children collecting everything their experience had taught them was fit to eat.[2] Their material possessions were scanty, and they huddled for shelter in caves along

[1] F. Dussart, "Les types de dessin parcellaire et leur répartition en Belgique," *Bulletin de la Société Belge d'Études Géographiques*, Vol. 30 (1961), 21–65.
[2] Christopher and Jacquetta Hawkes, *Prehistoric Britain* (Harmondsworth, Middlesex: Pelican Books, 1943).

the coast or near the chalk uplands where they could find the precious flint they needed for their stone tools. Perhaps these groups had their own "hunting grounds," and recognized the territorial rights of others, but the paleolithic people of Britain have left little trace upon the landscape.

The Neolithic Revolution, with its cultivated crops and domesticated animals, arrived late in Britain, brought in by successive waves of invaders from the continent of Europe.[3] The first invaders were simple people for whom the pastoral life was more important than farming. Apparently their womenfolk tilled tiny oval fields with digging stick or hoe, and harvested their meager crops of wheat and barley with sickles of sharpened flint. These people lived in clusters of small round huts on the open chalk uplands, but they too have left scanty traces of their presence upon the agricultural landscape.

The first enduring mark was made by the Celts, groups of farmers who arrived around 750 B.C. and began to build small isolated clusters of thatched huts and barns along the hill crests. The Celts used a light, bronze-tipped, ox-drawn plow to cultivate small square plots of half an acre or so, enclosed by banks of earth or stones. These plows dug a straight furrow and cut the sod, but they failed to turn it because they had no moldboards, and thus it was necessary to cross-plow the field at right angles to secure the desired tilth.[4] Cross-plowing produced squarish fields which still show clearly on aerial photographs; these are commonly called "Celtic fields," although they have no relationship to the modern patterns in any Celtic lands. Most of the fields which can still be seen are concentrated on the open chalk uplands, which have not been cultivated in subsequent millennia, but traces of such fields in other parts of England may have been obliterated by the activities of later plowmen.[5]

Shortly before the Roman invasion another group, the Belgae, arrived in England with iron tools and heavy plows drawn by teams of eight oxen. Unlike previous invaders, who had been more or less restricted to the light soils and open country of the chalk uplands, the Belgae began to attack some of the lowland forests, because they had iron axes with which to fell the trees, and plows heavy enough to till the heavier lowland soils. They plowed long furrows, to avoid having to turn the cumbrous teams of oxen more often than necessary, and their long narrow fields contrast sharply with the small squarish "Celtic" fields.[6] After the Roman invasion some of the native farmers prospered and adopted Roman ways of life. The Roman villas of England consisted of solidly built farm houses, barns, and outbuildings set around a courtyard amidst the large open fields of the farm. It is impossible to say how many were built by native farmers, but this was probably one

[3] Sonia Cole, *The Neolithic Revolution* (London: British Museum, 1961).
[4] M. S. Seebohm, *The Evolution of the English Farm*, rev. 2nd ed. (London: Allen & Unwin, 1952), p. 35.
[5] C. S. Orwin and C. S. Orwin, *The Open Fields*, 3rd ed. (London: Oxford University Press, 1967), p. 22.
[6] Seebohm, *The Evolution of the English Farm*, footnote 4, p. 43.

of those times in English history—the sixteenth and seventeenth centuries were another—when prospering farmers erected proud mansions to advertise their new-found success and rising status.

The most significant Roman contribution to the contemporary English rural landscape was not these villas, however, for our knowledge of them comes from the testimony of the archeologist's spade. Far more important for later generations was the construction of more than 5,000 miles of roads, many running straight as an arrow across the countryside.[7] The lines followed by many of these roads are still in use today, but of greater consequence is the fact that they provided ready-made routes whereby the later Anglo-Saxon invaders could penetrate wild and unsettled country more readily than if they had had to cut their own way through it.

The Celtic Fringe

Before examining the Anglo-Saxon invasion, however, it is appropriate to consider systems of land division and land tenure in the Celtic lands of the west. As each successive wave of warlike invaders from the continent descended upon the lowland plains of southern and eastern England, the unhappy native people fled farther into the hills of the north and west. These rugged uplands, swept by wind and lashed by rain, are hostile to man and beast. Their tough old Paleozoic rocks, stumps of ancient mountains, have been scraped bare by glacial ice or plastered by blankets of bog and acid, infertile, peaty soils; the small pockets of even moderately good soil are few and far between. This forbidding land became the refuge and the home of the Celtic people.

They evolved an infield-outfield agricultural system which was in marvelous harmony with the inhospitable environment; for the most part it has disappeared in modern times, but it can still be found in a few remote areas.[8] The infield was the best patch of ground available. It was enclosed by a stone wall or high earthen bank, received all of the manure, and remained in continuous cultivation. The standard crop was oats, which nourished both animals and men. The outfield, of lesser fertility, was essentially pasture land, but it was systematically cultivated under a system of "long ley." A patch of the outfield was enclosed by a temporary "fold" the summer before it was scheduled for cultivation, and cattle were penned on it to enrich it with their droppings. Then, after a crop of oats, or perhaps two, had been taken from it, the land was returned to grass until the time came to cultivate it once again.

---

[7] W. G. Hoskins, *The Making of the English Landscape* (London: Hodder and Stoughton, 1955), pp. 29–30. A technique called network analysis can be used to show that the centre of the Roman road network was near modern Birmingham, at the Britannica Inn, which must have sold the best mead in Britain; Peter Hutchinson, "Networks and Roman Roads: A Further Roman Network," *Area*, Vol. 4 (1972), 279–80.

[8] E. Estyn Evans, "The Ecology of Peasant Life in Western Europe," in William L. Thomas, ed., *Man's Role in Changing the Face of the Earth* (Chicago: University of Chicago Press, 1956), p. 229; and A. C. O'Dell and Kenneth Walton, *The Highlands and Islands of Scotland* (London: Thomas Nelson, 1962), pp. 108–11.

Each part of the outfield was cultivated in this fashion once every seven to ten years. Regular cultivation was necessary to renew the grazing quality of the pastures, which deteriorates fairly quickly under the excessive moisture conditions of the Celtic lands. Even today many farmers in Scotland still talk about the necessity of "taking the plow around the farm" in order to maintain the quality of their pastures.[9]

In many areas the uphill margin of the outfield, which was the upper limit of cultivated land, was enclosed by a stone wall, or head-dyke.[10] The improved grasslands below the head-dyke, and the rough grazings of heather and coarser grasses on the moorlands above it, were used as pastures for cattle and sheep. In addition, some animals were taken to distant upland pastures during the summer cropping season, which in Ireland traditionally began on St. Patrick's Day and ended on Hallowe'en.[11] The young folk, and sometimes entire families, left their winter homes and took their livestock and even their furniture to the summer pastures, where they lived in a hut known as a *booley* in Ireland, a *hafod* in Wales, a *shieling* in Scotland, and a *saeter* in Norway. Both the departure in spring and the return to the winter home in fall provided excuses for considerable festivities and ceremonials, which are still celebrated today even though most contemporary celebrants haven't the foggiest clue as to their original significance.

The most common type of winter home was a low, one-story, rectangular cottage which sheltered the members of the family in one end and their cattle in the other.[12] It was 12 to 15 feet wide, of varying length, and might be built of any locally available material—stone, timber, clay, wattle, even sod. The cottage was built on a slope at right angles to the contour, and the top end, with chimney, fireplace, and living quarters, was excavated into the hillside for warmth. The lower end housed the cattle, with the slope permitting refuse to drain away. During the winter months food for man and feed for beast both were often in short supply, and upon occasion a bowl of blood might be drained from one of the live animals to make a nourishing blood pudding for the family table.

In spring the cattle often had to be carried out to pasture because they were too weak to walk.

[9] John Fraser Hart, *The British Moorlands: A Problem in Land Utilization* (Athens, Ga.: University of Georgia Press, 1955), p. 39. The infield-outfield system had accustomed farmers in Scotland and Ireland to the notion of land rotation and long fallowing of exhausted land well before the first Scotch-Irish immigrants came to the United States, and this part of the cultural baggage of the Scotch-Irish settlers might have helped initiate the brush fallow system of land rotation which evolved in Appalachia; E. Estyn Evans, "The Scotch-Irish in the New World: An Atlantic Heritage," *Journal of the Royal Society of Antiquaries of Ireland*, Vol. 96 (1965), 39–49; and John Fraser Hart, "Abandonment of Farm Land on the Appalachian Fringes of Kentucky," *Mélanges de Géographie, Physique, Humaine, Économique, Appliquée, Offerts à M. Omer Tulippe* (Gembloux, Belgium: Editions J. Duculot, 1967), Vol. I, pp. 352–60.

[10] I. M. L. Robertson, "The Head-dyke: A Fundamental Line in Scottish Geography," *The Scottish Geographical Magazine*, Vol. 65 (1949), 6–20.

[11] E. Estyn Evans, *Irish Heritage: The Landscape, the People, and Their Work* (Dundalk, Ireland: W. Tempest, 1942), pp. 50–55.

[12] Evans, *Irish Heritage*, footnote 11, pp. 57–60.

The question of how these cottages were distributed has been debated with Celtic ferocity and intensity, and all the returns are still not in. Today the Atlantic coastal lands of Britain provide one of the best examples of dispersed farmsteads in Western Europe, but a great argument has raged over whether or not this has always been the case. Most combatants seem to agree that parts, at least, of the Celtic lands have had an infield-outfield system at one time or another, and that this has been associated with summer grazing of upland pastures. The principal questions at issue, it appears, are whether the infield was cultivated by one man, or by many; whether the rotational cultivation of the outfield was the work of one man, or a communal endeavor; and whether the pastures were grazed by the herds of many owners, or of one. In short, was the infield-outfield system, with transhumance, associated with a dispersed settlement pattern, or with one which was nucleated?

The only correct answer seems to be that it has been associated with both, because the Celtic lands have experienced more than one cycle of settlement dispersion, nucleation, and dispersion.[13] It has been suggested, for example, that invaders subjugated the native people and forced them to live as vassals in clustered settlements, whereas the invaders were free to live in isolated farmsteads. The growth of population, however, plus inheritance laws which required that a man's property must be divided equally among all of his sons, could have converted the isolated farmhouse of a father into a cluster of farmhouses inhabited by his sons or grandsons. This seems to be exactly what happened in Ireland, which once had many small, irregular clusters of ten to 20 farmhouses (known as *clachans*) in which nearly everyone had the same last name and, presumably, a common ancestor.[14] Conversely, a decline in population, such as was caused in Ireland by the Great Famine of 1845, could have enabled a single survivor to consolidate all of the land back into a single farm.

Patterns of land tenure on the infields of many clachans support the idea that the farmers of the clachan were all descendants of a single ancestor, and were farming as a group the land which he had once farmed himself. No farmer held a solid block of land, but all held plots (most often strips) scattered across the infield, as though a large field had been divided to give each person his fair share of each kind of land it contained, and then subdivided and redivided once again, often resulting in plots which were impossibly small. In one clachan 29 farmers had 422 plots of land; in another one man had 32 different plots; and in yet another no fewer than 26 people held shares in a half-acre plot.[15] Much of the good land was wasted in boundaries, and some farmers

---

[13] Sir John Clapham, *A Concise Economic History of Britain from the Earliest Times to A. D. 1750* (Cambridge, England: Cambridge University Press, 1949), p. 48; and V. B. Proudfoot, "Clachans in Ireland," *Gwerin*, Vol. 2 (1959), 110–22.

[14] Evans has suggested that nicknames are so common today in rural Ireland, Scotland, and Wales because they were necessary in clachans where everyone had the same surname; Evans, *Irish Heritage*, footnote 11, p. 48.

[15] Evans, *Irish Heritage*, footnote 11, p. 50.

had to walk hundreds of miles each year just going from one plot to another.

The evils of minuscule subdivision have long been apparent, and as early as 1800, or perhaps even before, attempts were made to consolidate the land into individual farms and solid blocks. In Ireland one result of the consolidation of land and the enclosure of the newly created fields by hedges or stone walls was the development of elongated strip farms, only one field wide, which ran from the valley bottom to the hilltop, so that each farm would have its share of each kind of land. These are often called "ladder farms," because the walls or hedges which separate individual farms look like the sidepieces of a ladder whose rungs are the cross hedges which divide each farm into individual fields.[16] The hard angularity of these new farms contrasts sharply with the gentle irregularity of the older infields and outfields.

Open Fields and Villages

"In the present-day rural landscapes of Western Europe," said Estyn Evans, "the contrast between the villages of the plain with their strip fields and the scattered farms of the Atlantic coast lands with their hedged fields is striking. And the attitudes of the folk who live and work in them are correspondingly different. The English village has lost many of its old functions and has changed its social structure, but there can be little doubt that it bred a disciplined tradition and a respect for law and order which, to the Irishman, appear to be the mark of simple minds."[17] The landscape of villages and strip fields, which is so characteristic of the lowlands of eastern England, owes much of its character to the consequences of the Anglo-Saxon invasion.[18]

The Anglo-Saxon invaders came from the marshy shores across the North Sea to a country that was essentially still a wilderness; within six centuries they had transformed it into a settled agricultural land. Few modern English villages can be traced back before about 450 A.D., when the Anglo-Saxon invasions began, yet the names of almost all of them can be found in the pages of the Domesday Book of 1086, which was compiled 20 years after the Norman Conquest.[19] The Anglo-Saxons divided the land into townships, which became the manors of feudal times, then the parishes of the church, and eventually evolved into the smallest civil divisions of contemporary England.[20]

The Anglo-Saxons came in small groups, often consisting of kinspeople. They were well-armed and pugnacious, and were constantly

[16] E. Estyn Evans, *Mourne County: Landscape and Life in South Down*, 2nd ed. (Dundalk, Ireland: Dundalgan Press, 1967), pp. 123–24.
[17] Evans, *Mourne County*, footnote 8, p. 234.
[18] Orwin and Orwin, *The Open Fields*, footnote 5, p. 23.
[19] H. C. Darby, "The Changing English Landscape," *The Geographical Journal*, Vol. 117 (1951), 379.
[20] Orwin and Orwin, *The Open Fields*, footnote 5, pp. 24–27.

clashing with other groups and with the natives.[21] When they were not fighting they worked together under their leaders to clear the forests and convert the land into large, open unfenced fields around their villages. Many of the villages were given the name of the leader of the group, plus a suffix such as -ing, -ham, -ton, -ingham, or -ington; for example, the name of the Leicestershire village of Peatling really means "Peotla's people," and Laxton in Nottinghamshire was originally Lexington, "the tun, or farm, of the people of Leaxa."[22]

The land of the village commonly consisted of a nucleus of cropland on the better soils, with meadows for hay along the stream or in the wetter parts, and woodlands or wastelands of bog, heath, and moor in the poorer marginal areas; rights to cropland were associated with rights to the meadow, woodland, and waste. Co-aration, or common cultivation of the cropland, was virtually necessitated by the fact that few men were wealthy enough to own a heavy plow and the eight oxen needed to pull it; one man might contribute the plow, another a team of oxen and so on. The standard unit of cultivation was the *strip* or the amount of land that could be plowed in one day. The strip was a "furrow long" (furlong), or the distance that a team of oxen could pull the plow without having to stop for rest. The ideal strip was 220 yards long and 22 yards wide, or an acre in size, but in practice the length and width varied with the nature of the soil and the lie of the land.[23] Although the ideal strip was seldom found, its dimensions still remain with us; the furlong length of an eighth of a mile is a measure of distance used in horse racing, and the width of 22 yards (or 66 feet) is exactly the length of a cricket pitch, and just a little more than the distance from the pitcher's mound to home plate in baseball.

A bundle of strips, containing few or many, but all running in the same direction, made up a more or less rectangular block known as a *furlong*; each furlong block had its own name, and may well have been the land that was cleared by the communal group in a single year.[24] The largest unit of all, the *field*, was a crazy quilt of many furlong blocks, of all sizes and shapes, at every sort of angle, containing hundreds of strips. The cropland of the village was commonly divided into three large open fields, each covering a few hundred acres and containing many furlong blocks, which were cultivated in the standard three-year rotation of food, feed, and fallow.

The slice of soil cut by the plow was turned toward the center of the strip when it was plowed, and in time this process built up a slight ridge down the center of the strip, and left a low furrow between each strip and those on either side of it; this furrow not only marked

[21] Clapham, *A Concise Economic History of Britain*, footnote 13, pp. 40–44.
[22] Hoskins, *The Making of the English Landscape*, footnote 7, p. 50.
[23] An acre originally was the amount of land that could be worked in one day; the origin of the word can be traced back to the Latin verb "agere," which means "to work."
[24] Maurice Beresford, *The Lost Villages of England* (New York: Philosophical Library, 1954), pp. 42–46; and Orwin and Orwin, *The Open Fields*, footnote 5, pp. 30–52.

the boundary between strips, but also facilitated drainage in a moist climate. Over the years this process built a slightly corduroy, or "ridge and furrow," surface, which can still be seen in many fields in the English Midlands.[25]

The strips which each man farmed were scattered far and wide over the cropland of the village. They were apportioned so that he had an approximately equal number in each of the three fields, and thus the only sensible, central place for him to live was in the village. Neither the strips nor the furlongs, and quite often not even the fields, were enclosed by fences of any kind, thus creating the historic landscape of farm villages amidst open fields which was common throughout medieval Europe.[26] By night the cattle and sheep were kept in their stalls, or in the fields in temporary enclosures known as *folds*; during the day they grazed the common wastelands, the fallow field, the meadows (after they had been mown for hay), and the barley stubble, after the cropland had been harvested in the fall. The lack of fences necessitated oxherds, cowherds, and shepherds (such as Little Boy Blue of the nursery rhyme) to keep the sheep out of the meadow and the cows out of the corn. Swine, which were only semi-domesticated, ran loose in the woods in care of a swineherd; it is not surprising that, in the medieval ballads, when a traveler encountered a wild man in the forest, he was almost invariably there tending swine.

The land of each village was originally separated from the land of other villages by empty wastelands of forest, bog, heath, and moorland, and thus it required no sharp delineation. As population increased, however, these wastelands had to be reclaimed and brought under cultivation to meet the growing demand for land, and it became ever more important to demarcate sharply the boundaries of the village as the territorial interests and designs of neighboring villages began to come into conflict.[27] The boundary was defined by prominent landscape features such as roads, brooks, stones, and trees (even, in one instance, "the beech tree wherefrom the thief was hung").[28] Young men were expected to learn the boundaries, or *marchos*, of their village, which were shown to them in an annual perambulation led by one of the village

[25] W. R. Mead, "Ridge and Furrow in Buckinghamshire," *Geographical Journal*, Vol. 120 (1954), 34–42; and M. J. Harrison, W. R. Mead, and D. J. Pannett, "A Midland Ridge-and-Furrow Map," *Geographical Journal*, Vol. 131 (1965), 366–69.

[26] For example, see R. A. French, "Field Patterns and the Three-Field System —The Case of Sixteenth-Century Lithuania," *Transactions and Papers, Institute of British Geographers*, No. 48 (1969), 121–34; *idem*, "The Three-Field System of Sixteenth-Century Lithuania," *The Agricultural History Review*, Vol. 18 (1970), 106–25.

[27] June A. Sheppard, "Pre-enclosure Field and Settlement Patterns in an English Township," *Geografiska Annaler*, Vol. 48, Ser. B (1966), 65; Hoskins, *The Making of the English Landscape*, footnote 7, pp. 216–24 has a lively discussion of the devastating effects of complicated property rights, and especially rights of common pasture, on the growth of towns in the nineteenth century; the influence of land division on the growth and character of one particular built-up area is examined in David Ward, "The Pre-Urban Cadaster and the Urban Pattern of Leeds," *Annals*, Association of American Geographers, Vol. 52 (1962), 150–66.

[28] Orwin and Orwin, *The Open Fields*, footnote 5, p. 27.

elders; the ceremony of "riding the marches" is still celebrated each June at Hawick, in Scotland, and perhaps elsewhere as well.

Clearance and reclamation continued intermittently through the centuries as population increased and declined. A piece of newly cleared land was called an "assart," from the French word *essarter*, which means to clear land of bushes and trees.[29] Many of the later assarts, especially those at the outer margins of the village lands, were not divided into the strips of the open field system, but were farmed as single units, and were known as "closes," because they had been enclosed from the waste.[30] Sometimes the newly reclaimed land was farmed from the village, but if it were too remote a completely new settlement, a "daughter village," might be created to house those who farmed it. The names of many modern villages reflect such a "mother-daughter" relationship; Much Tooting-on-the-Whistle, for example, would probably be the mother village of nearby Little Tooting-on-the-Whistle.

The Normans, Feudalism, and the Manor

Anglo-Saxon society recognized three classes of men: those of noble blood, or thanes; common free men, or churls; and slaves. The free men of the village maintained their lord by their labor, and in return expected leadership, protection, and justice. The average man had a lord of some sort to whom he gave presents, for whom he did jobs such as harvesting and woodcutting, and under whose leadership he fought. In addition to working on his own land, each man worked a certain number of days on the "demesne" (his lord's) and on the "glebe" (church) land. In time, virtually every activity in the village, every right and every duty (such as grazing rights, labor services which had to be performed, payments in kind which had to be made, etc.) came to be defined by a vast body of customary law, "the custom of the manor as interpreted in the lord's court."

The Anglo-Saxons combined communal cultivation with an acceptance of overlordship and the obligation of free men to cultivate the land of their lord. After the Norman Conquest William the Conqueror replaced native overlords with foreigners, his own men, and installed the feudal system firmly and systematically in every corner of his realm. Like a gangster chieftain after a successful raid, William divided up the loot among his henchmen; by 1086 not 1 percent of the land of England was held by the same men (or by their sons or widows) who had held it 20 years earlier at the time of the Conquest.[31] The feudal system which William installed was the standard form of political organization in medieval Europe, at a time when central government

[29] H. C. Darby, "The Clearing of the English Woodlands," *Geography*, Vol. 36 (1951), 74–75.
[30] Orwin and Orwin, *The Open Fields*, footnote 5, p. 104.
[31] H. C. Darby, "The Economic Geography of England, A. D. 1000–1250," in H. C. Darby, ed., *An Historical Geography of England Before A. D. 1800* (Cambridge, England: Cambridge University Press, 1951), p. 190.

was too weak to maintain effective law and order, and the economy was almost purely agricultural, based upon barter and exchange of services rather than on money. Under this system every man and every bit of land had an overlord; the man looked to his overlord for protection and justice, and expected to perform services for him in return. The duties of government devolved into the hands of a fighting aristocracy of hereditary landowners, the overlords, who could protect the peasants tilling the soil for them.

Under the feudal system no one actually owned land outright; he merely held it as a hereditary right from some overlord in return for the personal services he rendered to his overlord. In simplest theory, the king, as God's chief servant in his kingdom by divine right of birth, held all of the land of the kingdom, and parcelled it out among his nobles in return for the services, mainly military, which they were expected to render him. He assigned principalities to princes, duchies to dukes, marches to marquesses, earldoms to earls, counties to counts, and baronies to barons; each of these, in turn, might divide his land into smaller units, which were assigned, once again, in return for military services. At the very bottom of the list was the smallest unit of land, the manor, which was just large enough to provide one gentleman or knight (the lord of the manor) with a livelihood for himself and his family, plus the horse, armor, and weapons he needed to render military services to his lord and to protect the peasants who worked on his land.

In practice, of course, the feudal system was vastly more compli- cated than this, and it would have little interest but for the fact that the manor has left its mark indelibly upon the land. Each manor was an independent unit, and no two were exactly alike, yet they all had much in common. In its purest form the manor consisted of a village in the midst of unfenced open fields, and all of the land of the village (Fig. 3–1). Its most impressive structure was the home of the lord of the manor, or squire, which might have been a fortified castle in the early days, although later many castles were replaced by more comfortable mansions, or "halls." Close by the hall was the parish church and the home of the parish priest, for in many cases the church adopted the land of the manor as an ecclesiastical parish.

The main body of the village was the cluster of dwellings in which the peasants lived, and they all lived in the village; the manner in which these dwellings were arranged has received an inordinate amount of attention. The village had no stores, but there was a smithy where the blacksmith made nails, shoes for horses and oxen, and plow irons. Beside the stream was a water mill, where the villagers were supposed to have their corn ground. The lord of the manor claimed the right to make all of his tenants grind their grain at his mill, and required them to give him a certain portion of it for this service. This claim was a constant source of friction, because the landlord often leased these rights to a miller for a fixed rent, and millers tended to be crafty fellows, not above taking more than their fair share. Many of the villagers preferred to use the old-

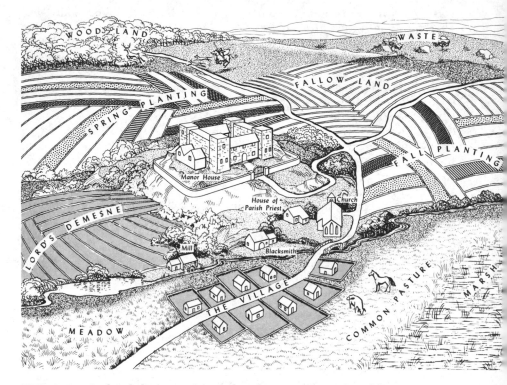

FIG. 3–1.   *Stylized diagram of a medieval manor. Reproduced by permission of the publisher from Wallace K. Ferguson and Geoffrey Bruun,* A Survey of European Civilization, *4th ed. (Boston: Houghton Mifflin, 1969), p. 155.*

fashioned hand mills, or stone querns, in which their womenfolk had been accustomed to grinding their grain at home.[32]

Although traces of the feudal past still linger on in such modern names as the Principality of Wales, the Duchies of Cornwall and Lancaster, the "Marches" of Wales and Scotland, and the ubiquitous "county" (which crossed the Atlantic Ocean to all 50 states except Louisiana and Alaska), the principal relict of feudal political organization is the basic unit of land-holding and decision-making, the manor. Over the centuries, in a very fuzzy fashion, the manor was largely superseded by the ecclesiastical parish for administrative purposes, perhaps in part because the poor laws were administered by the parish, and in part because the relationship between manor, parish, and village was seldom as neat and tidy as the simplistic picture I have painted here. The purist will argue, quite correctly, that the manor and the parish rarely encompassed pre-

----

[32] The exasperated Abbot of St. Albans, after half a century of disputation with his tenants, finally confiscated and broke all of their quern stones, and used the pieces to pave the floor of his parlor; Seebohm, *The Evolution of the English Farm*, footnote 4, p. 163.

cisely the same piece of ground, that a "manor" often failed to include an entire village or parish, and that many lords held manors *in* a place, not *of* that place. Nevertheless, village, parish, and manor were held together by the fact that the people of the village were expected to go, and for centuries did go, to the parish church, and the greater part of the village and the land of the parish probably was owned by a single squire, who was a key figure in decision-making as far as the use of the land was concerned. One might add, for the sake of the agricultural geographer, that the parish is the smallest unit of land for which agricultural statistics are available in Britain.

Laxton

Although most of the open fields were enclosed more than a century ago, and Lowland England has been converted into a countryside of hedgerows, a few museum pieces remain to give some notion of what the open fields once were like. Fragments of the older system have survived in several places, but the only open-field village which still remains is Laxton in Nottinghamshire, some 20 miles north and a bit east of the city of Nottingham.[33] In 1952 the manor of Laxton was purchased by the British Ministry of Agriculture, with the intention of maintaining it as a working example of the open-field system. This system of farming is completely obsolete, however, and although rents are low, the costs of moving animals, implements, seeds, manure, and crops are so great that the present generation of farmers is making only a modest living, and their children may well decide that they do not wish to make the sacrifices demanded of human exhibits in an open-air museum.

Laxton (originally Lexington, the tun, or farm, of the people of Leaxa) probably was settled by Anglo-Saxon farmers in the sixth or seventh century, which means that some of the strips in the modern open fields may have lasted more than a thousand years. The village is mentioned in Domesday Book, but the history of Laxton really begins in 1635, when the manor was sold to a London merchant, Sir William Courten, who commissioned a surveyor named Mark Pierce to make a detailed map showing each strip in the open fields and the name of the person who occupied it. The original copy of this splendid map is preserved in the Bodleian Library at Oxford; it is faithfully reproduced in the Orwins' *The Open Fields*.

In 1635 the strips of the demesne had already been consolidated into a solid block of land just north of the village (Fig. 3–2). The inner portions of the four great fields (West, Mill, East, and South), which presumably were the first areas to be placed under cultivation, were laid out in bundles of strips, or *furlongs*. They were cultivated in a three-year rotation in which West Field (318 acres) and East Field (134

[33] Orwin and Orwin, *The Open Fields*, footnote 5, pp. 61–192; for a briefer treatment, see J. D. Chambers, *Laxton: The Last English Open Field Village* (London: H. M. Stationery Office, 1964).

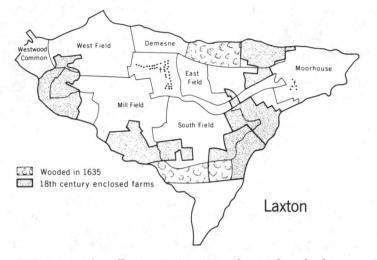

FIG. 3–2.   *The village of Laxton, 20 miles north and a bit east of Nottingham, England, where the traditional open-field system is still preserved. Compiled from maps in C. S. Orwin and C. S. Orwin,* The Open Fields, *3rd ed. (London: Oxford University Press, 1967), and J. D. Chambers,* Laxton: The Last English Open Field Village *(London: H. M. Stationery Office, 1964).*

acres) together were treated as a single rotational unit of approximately the same size as Mill Field (433 acres) and South Field (428 acres). The more distant areas, which must have been reclaimed later, consisted of solid blocks, or *closes,* except for Westwood Common, which provided all of the villagers with grazing land for their livestock when crops were in the ground. The Long Meadow, which was divided into narrow individual strips from which the villagers could cut their hay, occupied the wet, low-lying land along the brook which flows past the southern end of Laxton village toward the hamlet of Laxton Moorhouse, at the eastern end of the manor.

The "daughter village" of Laxton Moorhouse, with its own open fields, its own meadow land, and its own common land, was a miniature version of the mother village of Laxton. It even had its own "chapel," which is the standard English name for a place of worship subordinate to a parish church (in this instance, the parish church in Laxton village). By the thirteenth century the growing population in Laxton required more land than could conveniently be farmed from the village itself, and thus a part of the community was hived off to develop its own separate economic life in one of the remote corners of the manor. At about the same time the distant southern tip of the manor was made into a separate submanor. Much later, in the eighteenth century, some of the closes and wasteland at the margins of the manor were consolidated into solid blocks, withdrawn from the open fields, and made into five separate farms with their own new farmhouses and buildings.

Enclosure and the Agricultural Revolution

The medieval manor produced all it needed, but needed all it produced; the manorial system was essentially a closed, domestic, noncommercial economy. There was little surplus for trade, and the exchange of goods consisted largely of barter at the local level. Landownership conferred prestige, but it had little monetary value at a time when the rendering of services, rather than money, was the principal medium of exchange. As the economy gradually became more commercial, however, and as money began to circulate more freely, both landlords and their tenants were increasingly tempted to "commute" customary services into payments of money; the prosperous tenant might prefer to pay rent instead of having to work on his lord's land, and the landlord might be quite happy to convert customary services into cash income. The desires of landlord and tenant quite often failed to coincide, of course, and the long history of the demise of feudalism is the sorry tale of a series of wrangles between landlords who insisted upon customary services, or demanded money instead of services, and tenants who insisted upon commuting these services, or demanded their right to perform them instead of paying rent.

The two most dramatic episodes in the decline of feudalism and the open-field system in England both occurred in times of rural depopulation and rapid agricultural change, and both were periods of fairly widespread enclosure of open fields by hedges, stone walls, and other kinds of fences. The first wave of enclosures is associated with the Black Death, which first broke out in England in 1348, and wiped out a quarter to a third of the total population. The vastly reduced rural labor force was inadequate to maintain the traditional open-field cropping system just at a time when landlords were beginning to realize the profits which could be made from converting the open fields to pastures for sheep and cattle. In many villages the landlord evicted the dwindling band of surviving villagers, demolished their mud-walled hovels, and enclosed the land for grazing. Many deserted villages in Midland and eastern England date from this epoch.[34]

The first wave of enclosures excited fierce antagonisms against the enclosing landlords and great public sympathy for the evicted villagers, and some rather ineffectual laws were passed which forbade enclosure. Public policy changed, however, around 1660, in response to the innovations which triggered the Agricultural Revolution, and the second wave of enclosures began to gather momentum. It reached a peak between 1750 and 1850, when private Acts of Parliament approved the enclosure of some 4.5 million acres of land in nearly 3,000 parishes, mainly on the Midland Plain.[35] This second wave of enclosure coincided with the beginnings of the Industrial Revolution, for which it provided both food and

---

[34] Beresford, *The Lost Villages of England*, footnote 24, pp. 137–76.
[35] Hoskins, *The Making of the English Landscape*, footnote 7, pp. 138–39.

manpower; many farm workers pushed off the land by enclosure were pulled into the burgeoning urban industrial centers.

The second wave of enclosures was pushed forward vigorously by progressive landlords, whose ability to innovate was severely limited by the restrictions of the open-field system. In the open fields, for example, it was impossible to grow a crop such as clover, which would still be in the ground after the traditional harvest date, because the villagers forcefully defended their right of common grazing on the stubbles after that date. (The early settlers who brought this attitude with them to Massachusetts safeguarded their common lands so carefully, and in perpetuity, that reportedly it is still quite legal to graze a cow on Boston Common.)

The Agricultural Revolution of the eighteenth century was an upward spiral involving the enclosure of the open fields, the introduction of new crops, a new rotation system, and larger and better livestock. The new rotation of wheat, roots, barley, and clover produced more and better feed, which enabled the farmer to keep more and better livestock, and they no longer had to be half-starved during the winter months. On his newly enclosed pastures the farmer could improve his stock by controlled breeding practices which would have been impossible on the old open fields. The larger, more numerous, and better fed animals produced larger quantities of manure, which was reflected in higher yields of grain when it was returned to the fields.[36] Some crops were eaten directly in the fields by flocks of sheep which were penned inside temporary fences, or *folds;* this not only fertilized the soil and helped to maintain its fertility, but it also cut down on harvesting costs.

Not all of the land which was enclosed was used for crops, of course. Although many of the former strips of ridge and furrow have long since disappeared under the plow, in some areas enclosure removed the plow as well as the plowman, and since enclosure the land has been used mainly to pasture cattle and sheep. Over large parts of the English Midlands, for example, the old pattern of ridge and furrow is still clearly visible, especially when the sun hangs low in the sky and the shadows lie long across the ground.

The New Rural Landscape

Although traces of the older open-field system can still be seen in many areas, over much of Lowland England the enclosure movement and the Agricultural Revolution produced a new rural landscape, a patchwork quilt of small fields stitched together by hedgerows. The newly enclosed field was surrounded by a ditch, which marked its boundaries and facilitated drainage. Soil from the ditch was thrown up into a mound on the inner side, and this mound was planted with cuttings of hawthorn or other plants which would grow to produce a quickset (live) hedge. Ash or elm trees might be planted along the hedgerow to provide shelter,

[36] Eugene Mather and John Fraser Hart, "The Geography of Manure," *Land Economics,* Vol. 32 (1956), 25–38.

wood for domestic purposes, and aesthetic satisfaction. A good quickset hedge can last forever, because it can be *laid* or *pleached* to renew its stocktightness when it begins to grow old and thin. Each woody stem is half severed near its base, bent over into a horizontal position, and then woven around 4½-foot staves which have been driven firmly into the ground; the tops of the staves are woven together with twisted sprays of blackberry or willow. The stems quickly send up new shoots, put down adventitious roots, and soon the hedge is as good as it ever was; furthermore, a laid hedge provides a perfect jump for a field of fox-hunters hallooing across the countryside.

On many manors the first land to be enclosed was the demesne, especially if the lord of the manor had been able to make judicious trades and thus consolidate his open-field strips into a contiguous block of land; on many manors this was done long before enclosure. When customary services began to be commuted, and the lord could no longer depend upon them for the cultivation of his demesne, he might turn to hired wage labor, or he might rent the land to an ambitious tenant. Today the land of the old demesne is often called the "home farm" of the village, parish, or manor, and its operation is managed directly by the landlord or by his representative.

Awkward problems of farm layout were created when the land of a manor was enclosed and divided into individual farm units consisting of solid blocks of land, because all of the farm buildings were in the village. Quite often, therefore, the new farms were shaped rather like a slice of pie, with their points in the village where the necessary buildings were located (Fig. 3–3). A pie-shaped farm is quite difficult to operate, however, and eventually this difficulty was alleviated, at least in part, either by building a new "field barn" at the far end of the farm, or by building a completely new farmstead near the center of the farm when the older farmstead in the village began to deteriorate and require replacement.

The village remains the home of the landless farm laborer. Unlike North America, where land is cheap and labor is dear, in Europe land is so precious that farmers are prodigal of labor, which is relatively cheap, and six or eight men might be employed on a farm which could support no more than a father and son in the Middle West. In the United States the first adult male one meets on a farm is probably the farmer himself, but in England, where half a dozen men may be working in a field, the farmer is the man who still has his jacket on; the workers have taken theirs off and hung them on the hedge. The farm workers who live in the village form the lowest and largest group in a three-tiered rural social structure. Well above them is their employer, the farmer who manages the land, but the farmer himself is a tenant of the man who owns it.

Contrary to the beliefs of some rural sociologists, to whom any form of farm tenancy apparently is anathema, most tenant farmers in Lowland England are quite content with their status, and many of them would not even consider buying the land they are farming unless they are

An English Village more than a Century after Enclosure

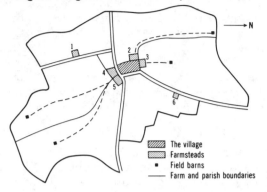

Ground plan of a British Barn

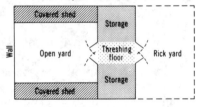

FIG. 3–3. The layout of farms in an English village, and the ground plan of a field barn. Before enclosure all farmsteads were in the village, and upon enclosure each farm was made wedge-shaped to provide access to its farmstead. Some farms have new farmsteads at more convenient locations, but others make do with field barns at the back end of the farm. The field barn is the standard English barn: rickyard in front where unthreshed sheaves of grain could be stacked in ricks, rectangular barn with large double doors in either side opening onto a central threshing floor, and stockyard behind where cattle could be fed for production of meat and manure.

literally forced to do so.[37] Before 1883 the English tenant farmer might have been at the mercy of his landlord, but a series of laws passed since that date have given him ever increasing freedom and security of tenure, until today he can be removed from his tenancy only if he can be convicted of very bad farming practices before a jury of his fellow farmers, and they are understandably reluctant to pass judgment on one of their own. His rent, which might have been set some years ago, is unrealistically low, and on many estates the rents do not even cover the costs of taxes, upkeep, and management, much less provide any return on the capital invested in the land.

The landlord has the prestige of landownership, of course, and he

[37] Henry C. Taylor, "Food and Farm Land in Britain," *Land Economics*, Vol. 31 (1955), 24–34.

may be reluctant to rupture a patriarchal relationship with his tenants by haggling over rents. When a tenancy falls vacant, however, he may prefer to sell the land rather than taking on a new tenant, and often a farm must be sold when a landlord dies and heavy death taxes are levied against his estate. The sitting tenant commonly is given first refusal, and usually buys the land, but he may have to borrow part of the purchase price. Purchase of the land removes part of his working capital, and if too much of his money is tied up in land and buildings he may be seriously short of the money he needs for the operation and maintenance of the farm. In short, many tenant farmers in England are happy to be tenants, and have no desire to become landowners.

Land Division Among the Boers

Property boundaries in Europe evolved over the centuries, as notions about property rights developed, and many boundary lines were marked by man made features or, in places, only by familiarity and tradition. The people who went out from Europe to settle distant lands carried with them firm beliefs about the importance of property rights, but only the vaguest of ideas as to how property boundaries might best be established and marked, and this created some problems.

The Boer Voortrekkers of South Africa were stockmen who followed their cattle into the dry interior of the continent after the British outlawed slavery in their original colony at the Cape of Good Hope. The Boers laid out large holdings of 6,000 acres or more. Each man selected a piece of land he liked, and drew the boundaries as he saw fit. In the early days a settler might define his boundaries by riding a horse for half an hour, at a walking pace, in several directions from a central point

Original survey lines near Swellendam, Cape Colony

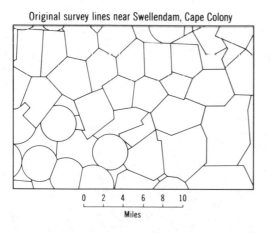

FIG. 3–4. *Original survey lines near Swellendam, Cape Colony, South Africa. Redrafted, with the author's permission, from A. J. Christopher, "Natal: A Study in Colonial Land Settlement," unpublished doctoral dissertation, University of Natal, 1969.*

("Ordonnatie"), which was often a conspicuous landmark, or a spring near which the settler planned to build his home. He could not claim land which had already been chosen by a neighbor, and so it was customary to locate the Ordonnatie at least an hour's ride from any neighboring one. This system produced roughly circular landholdings, which did not completely fill the land, and sons or latecomers might claim the "gores" which lay between them (Fig. 3–4).

Straight lines between convenient points eventually replaced the awkward circular boundaries. The most popular location for boundaries was a ridge crest, where straight lines could be drawn between beacons erected at prominent points. Most properties had irregular shapes, because the best land was selected first. Gores between properties are still reflected in contemporary fields, many of which have at least one "corner" which is a remarkably acute angle. The large original holdings were soon divided into smaller properties, because the Roman-Dutch tradition of Boer law required that the land be divided equally among all heirs, but the subdivisions retained the irregular shape of the original holdings. Almost half of all modern farm boundaries still follow the boundary lines of the original landholdings, despite a century or more of subdivisions and amalgamations.

CHAPTER 4   *land division*
in America

The colonists and settlers who came from the Old World to the New came from countries and areas where ideas about land division and land ownership had been evolving for centuries, or even millennia. In medieval Europe no man and no land was without a master; few men owned land in fee simple (without limitations or restrictions as to whom it could be sold, given, or bequeathed), but property rights were paramount, and they were jealously guarded, because title to land was the principal source of wealth and prestige. The people who emigrated from the Old World were in general agreement as to the desirability of property rights, but they differed considerably on details, and they also differed in their ability to adjust their preconceived notions to the unlimited amounts of land available in the New World. The Spanish, the French, the Dutch, and the British, each brought with them their own ideas as to how the land should be divided up and used.

They descended upon people whose notions of property rights to land were quite simple. Although some tribes grew corn, beans, melons, tobacco, and other crops, often in communal plots, for the most part Indian agriculture was a rather desultory enterprise which merely supplemented hunting, fishing, and gathering from the wild. Tribes had their traditional hunting grounds, but constant bickering between neighboring tribes indicates that their boundaries were flexible.[1] Even today the tidy-minded white bureaucrats on the Navajo Indian Reservation in northeastern Arizona, who would like to be able to draw neat property

---

[1] One exception was the boundary between the hunting grounds of the Bayougoulas and the Houmas, which was marked by a large red pole on the bank of the Mississippi River. The French name for this red pole, Baton Rouge, is preserved in the name of the capital of Louisiana; F. J. Marschner, *Land Use and Its Patterns in the United States*, Agriculture Handbook No. 153 (Washington, D.C.: Department of Agriculture, 1959), pp. 9–10.

lines on their maps, are frustrated by the ease and rapidity with which the boundaries of individual grazing areas can fluctuate.

The early white settlers in North America, steeped in the European tradition, simply could not understand the Indian attitude toward land. And vice versa. When a group of strangers, dressed in outlandish garb, approached an Indian chief and offered him a collection of beads, shells, mirrors, knives, and other trinkets in return for his land, the chief probably reacted much as a contemporary Manhattanite would react if the Past Imperial Potentate of the Grand and Majestic Lodge of Keokuk, clad in full regalia, were to stop him on the street and offer to swap him a thousand bushels of corn, a hundred bales of hay, and a good-field no-hit second baseman for his share of the Brooklyn Bridge.

The Spanish and the French, for the most part, did not confuse the Indians with such niceties; they simply took what they wanted and claimed title to it by right of conquest. The Dutch and the English were a bit more fastidious in paying for what they took (every school child knows the story of Peter Minuit's purchase of Manhattan Island for 24 dollars), but they accomplished almost as little when they treated the astonished Indians as high contracting parties to transactions which the Indians could not possibly comprehend. The British notion that Indian tribes were equivalent to sovereign states, with whom formal "international" treaties could be negotiated, continued to bedevil the early years of the youthful United States as the nation expanded westward.

The Spanish

The common American view of the Atlantic Seaboard as "colonial America" is a bit myopic, because some of the earliest settlements were founded along the southern margin of the United States, from California to Florida, which was once part of the Spanish Empire in the New World. This was one of the most remote and least important outposts of empire, however, and the rather limited expansion which occurred along it was motivated largely by religious or defensive considerations: either to convert the infidel Indians, or to prevent hostile groups from occupying territories claimed by Spain. Florida, for example, was a Spanish possession from 1566 to 1763 (197 years, a period equivalent to the time between the signing of the Declaration of Independence and 1973), but St. Augustine was the only place which the Spanish held continuously, and modern Florida shows few traces of the Spanish occupation.[2] The land was officially regarded as the property of the Indians, and individual Spaniards were not permitted to acquire title to it, which rather effectively forestalled widespread settlement.

Fears of foreign aggression inspired many of the territorial activities of the Spanish (and later, Mexican) governments farther west. Large tracts of land were granted along the frontier, both to get the country settled and to protect it against encroachment: by the French and

2 Ralph H. Brown, *Historical Geography of the United States* (New York: Harcourt Brace Jovanovich, 1948), pp. 68–76.

"Yankees" in Texas, by raiding Indian tribesmen and Yankees in New Mexico and southern Colorado, and by the English, Dutch, French, and Russians in California. Four kinds of grants were made: (1) common grants to Indian pueblos, which did little more than confirm the rights of the Indians to the land they were already using; (2) individual grants to such influential persons as government officials, army officers, and men of wealth, (3) community grants to groups of settlers; and (4) development grants to impresarios, who were required by the terms of the grant to bring in settlers.[3] The last three types were controlled, in the early Spanish days, at least, by rigid and comprehensive laws stipulating that they must be in unoccupied areas which could be used without injury to the Indians, and with their voluntary consent.[4]

In Texas the Mexican government continued the Spanish tradition of granting land in large blocks; an impresario received one square league (approximately 4,428 acres, or roughly seven square miles) for each family he brought in to settle on it.[5] The grant made to Stephen F. Austin, which embraced the better part of more than a dozen contemporary Texas counties, was larger than Connecticut and Rhode Island combined. After independence the Republic of Texas granted the settler a stipulated acreage of land which he could stake out on any part of the unappropriated domain, without reference to any system of land survey.

Land division was more closely regulated in the Spanish-Mexican settlements of the upper Rio Grande valley north of Santa Fe, where colonization was based on farming villages, or *pueblos*, in irrigable valleys. Strict rules governed the layout of the pueblo, which centered on a plaza. *Solares*, or building lots, were provided within the pueblo itself; *suertes*, or strips of agricultural land, were laid out at right angles to the stream used for irrigation; and an *ejido*, or common pasture, was established near the pueblo.[6] These features are still clearly visible on aerial photographs of the modern village of San Luis in central southern Colorado.[7] The *ejido*, an area of 900 acres of unfenced meadow land on which each family had the right to graze ten head of cattle or horses, is south and a bit east of the village. Farther upstream are the *suertes*, most of which are only 100 to 220 feet wide, but extend ten to 15 miles back to the watershed which separates this valley from the next one. The back parts of these strips are in dry range land of such slight grazing value that no one really knows or cares much about the location of boundaries within it.

[3] Allan G. Harper, Andrew R. Cordova, and Kalervo Oberg, *Man and Resources in the Middle Rio Grande Valley* (Albuquerque: University of New Mexico Press, 1943), p. 18.

[4] Ralph E. Twitchell, *Spanish Colonization in New Mexico in the Onate and De Vargas Periods*, Number 22 (Santa Fe: Historical Society of New Mexico, 1919), p. 5.

[5] D. W. Meinig, *Imperial Texas: An Interpretive Essay in Cultural Geography* (Austin: University of Texas Press, 1969), pp. 29 and 45; and John Whitling Hall, "Sitios in Northwestern Louisiana," *Northwest Louisiana Historical Association Journal*, Vol. 1, No. 3 (Spring 1970), 3 and 7.

[6] Harper *et al.*, *Man and Resources*, footnote 3, p. 18.

[7] Alvar W. Carlson, "Rural Settlement Patterns in the San Luis Valley: A Comparative Study," *The Colorado Magazine*, Vol. 44 (1967), 111–28.

Occupancy and use, and use based upon water for irrigation, were the keys to the Spanish-Mexican land system in California, where land was plentiful but water and people were scarce. The Spanish made fewer and larger grants, but even the Mexican grants were large; the minimal grant under Mexican law was a square league (about seven square miles), and the legal maximum, which apparently could be exceeded without too much difficulty, was one square league of irrigable land, four square leagues of land which could be used for dryland farming, and six of grazing land, for a total of 11 square leagues, an area slightly larger than the District of Columbia. The boundaries of land grants in California seem to have been located wherever the grantee wanted to put them, and they were surveyed, if at all, in the most offhand manner; gaps or overlaps in marginal areas created no problems, because such areas were used only extensively, at very best.

The headaches began when Americans started to pour in and demand clear title to the land on which they settled. The land grants in California which were eventually confirmed by the United States were concentrated in just those sections of the state which have attracted the largest numbers of people: the areas around San Francisco Bay, Monterey Bay, Santa Barbara, the Los Angeles Basin, and a scatter in the Sacramento Valley.[8] Many problems beset a new settler in trying to establish title to his land, and the settlement of some areas was considerably delayed because newcomers were unwilling to file for land which might be taken away from them after they had improved it.[9] Large land grants dating from the colonial period have strongly influenced urban growth in California even in recent years, especially around the peripheries of the Los Angeles Basin. The new technology, which permits residential development of hilly land previously good only for grazing, requires large blocks of land, and it is easier for developers to acquire such blocks from old land grants which have never been divided than it is to try to piece together a block of suitable size by acquiring many small separate tracts.

The French

The French government tried to export a vestigial form of feudalism, the seigneurial system, to its possessions in Canada. Large acreages of land, or *seigneuries*, were granted to *seigneurs*, who were supposed to ensure that the land was settled. In return, the seigneur had a variety of feudal rights over his tenants, or *censitaires*, such as the right to receive token rent payments from them, to require them to use his mill for grinding their grain, and to demand various work services from them.[10] Harris has concluded that the feudal elements of the system were largely irrele-

[8] Ida May Shrode, "Early Settlement of California and the Southwest," in Clifford M. Zierer, ed., *California and the Southwest* (New York: John Wiley, 1956), p. 115.
[9] Henry George, *Our Land and Land Policy* (New York: Doubleday, 1911), pp. 39–40.
[10] Marcel Trudel, *The Seigneurial Regime*, Booklet No. 6 (Ottawa: Canadian Historical Association, 1963), pp. 10–13.

vant to the life of early French Canada, although later some French Canadians may have tried to maintain them as a defense against English culture after the conquest of 1764.[11] The system of land division associated with the seigneurial system, however, remains vividly imprinted upon the landscape of rural Quebec.[12]

The seigneuries were laid out, for the most part, at right angles to the St. Lawrence River, the principal axis of French Canada. They had a relatively narrow frontage on the river, but extended far back into the interior. The land of the seigneury, in turn, was divided into long, narrow strips known as *rotures*. The average roture had a river frontage of two to three arpents (384 to 576 feet), and was commonly ten times as long as its width. A straight line parallel to the general course of the river marked the back end of the first "range" of rotures; this line was often followed by a road which provided frontage for a second range after the first had been occupied.

The "long-lot" system of land division was cheap and easy to survey, and it commonly provided each holding with a variety of vegetation and soil types on several different slopes, terraces, and kinds of rock. It gave frontage on a transport artery (whether river or road) to a maximum number of holdings, and along a river it provided maximum access for fishing. Each family could live on its own land, but still be close to neighbors. The land could be placed in cultivation progressively, by clearing the forests or draining the marshes nearest the farmstead in the beginning, and leaving the more remote parts of the holding until later. On the negative side, the twists and bends of a meandering stream could complicate the task of the surveyor, and a change in the river's course after the survey could lead to considerable litigation. And perhaps most critical of all, some painfully narrow farms were produced when a censitaire's land was divided equally among his children, especially if they were as numerous as they often were in French Canada.

Despite its drawbacks, the French carried this system with them wherever they settled in North America. They were more interested in fur trading than in farming, to be sure, and apart from the lower Mississippi valley, the settlements which developed beside their trading posts were small and grew slowly. The principal mark they have left upon the land is the remnants of their land division system (Fig. 4-1).[13]

[11] Richard Colebrook Harris, *The Seigneurial System in Early Canada: A Geographical Study* (Madison: University of Wisconsin Press, 1966), pp. 7–8 and 193–97.

[12] Peter B. Clibbon and Jacques Gagnon, "L'évolution récente de l'utilisation du sol sur la rive nord du Saint-Laurent entre Québec et Montréal," *Cahiers de géographie de Québec*, 10th year, No. 19 (1965–66), 55–72.

[13] Although the topographic maps of Louisiana are better known, one of the best illustrations of the French survey system, complete with long-lots and common grazing area, is the Vincennes, Ind.-Ill., 1:62,500 quadrangle (1913 edition); the history of the area is discussed in R. Louis Gentilcore, "Vincennes and French Settlement in the Old Northwest," *Annals*, Association of American Geographers, Vol. 47 (1957), 285–97. Traces of long-lots can also be seen on the Dearborn, Mich., 1:24,000 quadrangle (1952 edition), the De Pere, Wisc., 1:62,500 quadrangle (1954 edition), the Renault, Ill.-Mo., 1:62,500 quadrangle (1911 edition), and the St. Paul, Oreg., 1:24,000 quadrangle (1970 photorevision). Russel L. Gerlach has compiled a map of the Spanish rectangular land grant at Old Mines in Washington

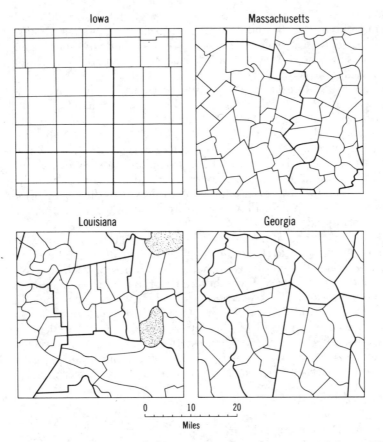

FIG. 4–1. *Minor civil divisions in selected portions of Iowa, Massachusetts, Louisiana, and Georgia, showing the effects of different land survey systems.*

The Dutch

The early Dutch settlements in the Hudson valley were yet another attempt to export the feudal tradition to the New World, "though it became famous more for what it attempted than for what it accomplished." [14] A *patroon* was granted 16 miles on one side of a navigable river, or eight miles on both sides, and as far back as he could occupy the land, on condition that he establish a colony of at least 50 adults within four years; he could select any stretch of the river which pleased him,

County, Missouri, which shows quite clearly that it was divided into long-lots by French miners. Stephen F. Austin, who lived in the French settlement of Ste. Genevieve, Mo., as a boy, carried the idea of the long-lot to Texas; Terry G. Jordan, "Antecedents of the Long-Lot in Texas," *Annals,* Association of American Geographers, Vol. 64 (1974), 70-86.

[14] D. W. Meinig, "The Colonial Period, 1609–1775," in John H. Thompson, editor, *Geography of New York State* (Syracuse, N.Y.: Syracuse University Press, 1966), p. 125.

and was permitted to move to a new site if he changed his mind about his first location. His tenants had to pay him a tenth of their farm products, plus a cash rent, and they were obliged to perform services such as wood cutting, hauling, and the repair of roads and buildings.

"In a new and promising country where the natives were friendly, the transportation easy, the land fertile, and all other conditions favorable," wrote Maud Wilder Goodwin, this attempt to transplant the feudal system produced "a sparse and sullen tenantry, an obsequious and careworn agent, an impatient company, and a bewildered government." [15] In 1638 the patroonship system was abandoned in favor of smaller grants to individuals, but few Dutch settlers saw fit to forsake the comforts of their homeland to accept them. After the English takeover of 1664 the royal governors reverted to a policy of making large grants to favored individuals, often for speculation rather than colonization. The large blocks of privately held land along the lower Hudson River were a fairly effective barrier to the settlement of central and western New York until westward-moving New Englanders swept around them to the north.

The English

NEW ENGLAND. Apart from the Dutch settlements in the lower Hudson valley and some inconsequential efforts by Swedes on the Delaware, Englishmen set the policies for land division along most of the Atlantic Seaboard of the United States. [16] The very name of New England is appropriate, because most of the settlers actually did come from England, at least in the early years, whereas areas south of the Hudson were settled by a more diversified group of people drawn from many parts of northern and western Europe. The early New Englanders were a fairly stiff and intolerant lot who brooked no nonsense, a fact which shows up quite clearly in their system of land division and settlement.

Before 1629 a few tracts of land in New England were granted to individuals, but for almost a century thereafter land was granted only to groups of worthy men, or proprietors. The basic unit was the township, or "town," a more or less compact block of land whose ideal size was six miles square. Many towns are bounded by straight lines, although few were actually square (Fig. 4-1). New grants were located close to towns which had already been settled, to ensure that settlement advanced inland in serried tiers, but gores of "no man's land" were sometimes left between old and new towns when grants were made.

Near the center of the town the proprietors laid out a village with a meeting-house (which served as both church and town hall) and the

[15] Maud Wilder Goodwin, *Dutch and English on the Hudson* (New Haven, Conn.: Yale University Press, 1919), p. 46.

[16] Glenn T. Trewartha, "Types of Rural Settlement in Colonial America," *Geographical Review*, Vol. 36 (1946), 568–96; Charles O. Paullin, *Atlas of the Historical Geography of the United States*, ed. by John K. Wright, and published jointly by the Carnegie Institution of Washington and the American Geographical Society of New York in 1932, Plates 40–56; and Marshall Harris, *The Origin of the Land Tenure System in the United States* (Ames: Iowa State College Press, 1953).

home lots of early settlers fronting on a town common, or village green, of an acre or two. Crops were grown on large fields of several hundred acres. The fields were located in different directions and at different distances from the village. Each field and each meadow was divided into strips and parcels of varying size. Each man's farm consisted of a scatter of such strips and parcels, their number and size depending on the size of his family, the wealth and property he had brought with him, or some other measure of his value to the community. Land not needed at first was held in common for future allocation as population increased or new settlers joined the community.[17]

When a town was fully settled and more land was needed, a group of its members would live off and obtain a new block of land elsewhere. The repetition of this process produced the characteristic New England pattern of nucleated villages and fragmented farms. Satellite settlements which developed in the more remote corners of some of the larger towns often adopted some modified form of the town name; the town of Eden, for example, might boast places with such names as Eden Center, Eden Corners, and Eden Junction, or less imaginatively, North, South, East, or West Eden.

The nucleated village associated with the division of the land into compact towns was essential to the social, educational, and religious well-being of the stern New Englander, who was more interested in preserving a way of life than in amassing this world's goods; many towns, in fact, forbade the sale of land to outsiders without the approval of the town meeting. The New Englander was just as concerned about the morals of his neighbor as he was about his own, and it obviously was more difficult for either to be corrupted when they were living side by side in constant vigilance. The man who lived on an isolated farm was much more easily tempted to develop disrespect for authority, use strong language and even stronger drink, and even go so far as to accept bad weather and poor roads as an excuse for failing to attend church on the Sabbath.

The original New England system began to break down under pressure from men with less interest in God and more in Mammon. Powell's history of local government in one Massachusetts town, Sudbury, describes acrimonious quarrels over the division of land which erupted less than two decades after the town had been founded.[18] Eventually the dissidents, mainly a group of landless young sons, obtained their own grant and formed a new town. Toward the end of the seventeenth century large blocks of land on the frontier were being sold to men of wealth or prominence, who never intended to settle the land themselves, but planned to subdivide it and sell it at a profit to individual settlers. In time even the states came to look on the sale of their western lands as a source of revenue.

[17] David Lowenthal, "The Common and Undivided Lands of Nantucket," *Geographical Review*, Vol. 46 (1956), 399–403.
[18] Sumner Chilton Powell, *Puritan Village: The Formation of a New England Town* (Garden City, N.Y.: Doubleday Anchor Books, 1965).

PENNSYLVANIA. King Charles II, after his restoration to the English throne in 1660, paid off some of his political debts by granting huge tracts of land in the New World to "proprietors," who were authorized to dispose of these vast private estates as they saw fit. Different proprietors, quite naturally, had different notions as to how they should dispose of their land. The proprietors in Pennsylvania, Delaware, and North Carolina sold their land in small blocks, but in Maryland, Virginia and South Carolina the proprietors granted large estates, manors, or plantations. In New Jersey the "true and absolute lords proprietors of all the Province," Sir George Carteret and Lord John Berkeley (or their heirs), sold their grant intact; within two decades it had been divided into two parts, had more than two dozen proprietors, and had received a legacy of confused land titles which required almost a century of litigation to unravel.

The most successful proprietor was the gentle genius, William Penn, who in 1681 was given absolute title to all land between the fortieth and the forty-third parallels, and five degrees of longitude west of the Delaware River.[19] Penn advertised widely, and sold land to all comers. He planned to lay out the land in neat squares to encourage community settlement, but this idea failed to appeal to individual buyers, and much of the colony was settled without any regular system of land survey.

Penn's colony became one of the most important culture hearths in the United States. Southeastern Pennsylvania contains the only large area of really good farmland east of the Appalachian Uplands, and it supported a prosperous agricultural economy, thriving small towns, and the bustling seaport of Philadelphia. Before the Revolutionary War this city was the principal port of immigration to the American colonies. Many stolid, industrious Germans, later to become known as the "Pennsylvania Dutch," settled west of Philadelphia and applied their expertise to the fertile soil or in small factories. The restless, cantankerous Ulster Scots pressed on westward to the frontier, but in passing through the German settlements they learned many skills (such as farming the land or building log cabins), and acquired tools and implements (such as the "Kentucky" long rifle or the Conestoga wagon) which they used in taming the western wilderness.

One of William Penn's own most enduring monuments is the widely copied plan for his city of Philadelphia, which had a grid pattern of streets named in an orderly fashion.[20] The east-west streets running back from the Delaware River, which were named after trees and flowers, make right angle intersections with the north-south streets parallelling it, which were numbered consecutively away from the river. The two princi-

[19] A dispute with Penn's fellow proprietor, Lord Baltimore of Maryland, over the location of the southern boundary was not finally settled until 1767, when Charles Mason and Jeremiah Dixon were brought from England to survey the line at 39°43′26″ which marks the traditional boundary between the North and the South, the Mason and Dixon Line.

[20] Is is hard for a contemporary American to realize just how impressive a grid pattern of streets must have been to anyone who knew only the "cowpath" street patterns of cities in Europe.

pal streets, Broad and High (now Market), meet at a central square formed of rectangular corners cut out of the four adjoining blocks. The City Hall occupies this square in Philadelphia, but it was graced by the county courthouse in many other early cities of southeastern Pennsylvania which adopted the "Philadelphia plan." [21] This plan, including the central square and the system of street names, was emulated in many cities of the interior.

THE SOUTHERN COLONIES.   The systems of land division and settlement in Pennsylvania and the colonies to the south were far more discriminating than the practices of New England, but the sobriquet of "indiscriminate location" was hung upon them (perhaps by green-eyed New Englanders?), and it has stuck. In New England land was granted in solid blocks close to areas that had already been settled, which ensured that settlement advanced inland in compact tiers, but forced settlers to accept the poorest land along with the best. Settlers in the southern colonies labored under no such constraints; each man was permitted to select the unclaimed land which pleased him best, and the poorer land was bypassed. The result, of course, was an extremely irregular and unsystematic pattern of land division, and one which could produce conflicting claims and protracted litigation over titles unless the land was surveyed with a degree of skill which few colonial surveyors possessed (Fig. 4–1).

In the southern colonies the prospective settler obtained from the authorities a warrant which stipulated the number of acres of land he was entitled to claim and settle, but did not specify their location. He then proceeded to select a likely looking tract of unappropriated land, preferably close to water and far enough from other settlers to provide a reasonable degree of assurance against claims of encroachment. After he had laid out his land, he had it surveyed, and presented a copy of the survey, together with his warrant, to the proper county official or land office. After a specified period of time, if no valid protest were raised, he was issued a patent to the land, which was the equivalent of a deed in fee simple. This patent, of course, could be no better than the survey on which it was based, and colonial surveyors (*pace* George Washington) were notorious for their inexperience, bungling, and incompetence (Mason and Dixon finally had to be brought from England to survey the boundary between Pennsylvania and Maryland properly).

Inaccurate surveys with poorly marked boundaries combined with loose administration of grants and titles to provide a fruitful source of

[21] Edward T. Price, "The Central Courthouse Square in the American County Seat," *Geographical Review*, Vol. 58 (1968), 29–60. Changing fads and fancies in the division of urban land is an intriguing but relatively neglected theme in cultural geography. Price's study, and Dan Stanislawski, "The Origin and Spread of the Grid-Pattern Town," *Geographical Review*, Vol. 36 (1946), 105–20, are so fascinating that one wonders why the subject has attracted so little attention. Furthermore, the layout of later residential subdivisions has been almost completely ignored by geographers, to the best of my knowledge, despite the remarkable diversity and variety which are displayed on detailed maps and aerial photographs of almost any city. In considerable degree, I suspect, these differences reflect the era in which that part of the city was subdivided; how if at all, have they influenced subsequent urban life?

litigation over land; the entire state of New Jersey could have been bought for less money than was spent in lawsuits over the ownership of its land. Many of the lines and boundary markers used in a "metes and bounds" type survey were ridge lines, streams, or such temporary features as trees or large rocks. Some boundaries were never carefully surveyed; even today, many county lines on detailed topographic maps of parts of Kentucky carry the notation "Indefinite Boundary," and it is estimated that no taxes whatsoever are paid on a considerable acreage of land in that state, because the owner tells the officials of each county that his land is located in the other.

With an empty wilderness stretching apparently without end into the interior, the colonies and later the states were only too happy to pick up a bit of change by selling warrants for land, or to use warrants instead of cash to pay off their financial and military obligations. Speculators were not above selling the same piece of land to several different purchasers, and some officials were little better; land grants totalling 29,097,866 acres were made in 24 Georgia counties whose total area was only 8,717,960 acres.[22] Furthermore, a man with no legal title to land could squat on it and claim preemption rights on the basis of occupancy and development; in 1779 the Kentucky land court allowed anyone who had settled on the land to claim 400 acres, with an additional 1,000 acres if he had improved it.

The Americans

THE OLD NORTHWEST. Problems of conflicting land claims were so serious in the seaboard colonies, and were becoming so threatening west of the Appalachians, that the founding fathers determined they must be solved before settlement could be allowed on the western waters. Their largest block of unclaimed land was the Northwest Territory, the area north of the Ohio River, east of the Mississippi, and south of the Great Lakes, which the Virginia militia had captured from the British during the Revolutionary War. This Old Northwest became the nucleus of the original public domain of the United States when the Commonwealth of Virginia ceded it to Congress in 1784; other states relinquished some rather extravagant claims to this territory, and to western lands farther south, which were also made part of the original public domain.[23]

A Congressional committee was appointed, under the chairmanship of Thomas Jefferson, to prepare an overall plan for the systematic survey and subdivision of the public domain, which would precede its alienation. The committee's proposal, as modified and adopted by Congress in the

[22] Marschner, *Land Use,* footnote 1, p. 14.
[23] Virginia reserved the right to enough good land in the Northwest to redeem the warrants (ranging from 100 acres for a private with less than three years service to 15,000 acres for a major general with four) with which she had rewarded her Revolutionary War veterans. She selected a tract between the Scioto and Little Miami Rivers in Ohio (now known as the Virginia Military District) and a small block of land in southern Indiana at the Falls of the Ohio River across from Louisville, Kentucky.

Ordinance of 1785, is variously known as the Congressional, General Land Office, National Land, Public Land, and Township and Range Survey System (Fig. 4–1). The fundamental features of the system, whose details are belabored in many introductory geography textbooks, are: (1) land had to be surveyed before it could be alienated and settled; (2) the survey lines should be oriented in cardinal directions; (3) the land should be divided into townships six miles square, which were subdivided into sections one mile square (or 640 acres).[24]

The state of Ohio was the testing ground where experiments were made with several different kinds of rectangular land survey systems. The township and range form which was used in northwestern Ohio, the last part of the state to be surveyed, became the basic system which was subsequently extended to most parts of the United States west of the Appalachians. In many places the straight lines of the survey are followed by the boundaries of states, counties, and minor civil divisions within counties. The survey is also responsible for the fact that many American farmers reckon their land in terms of sections (640 acres), quarter sections (160 acres), and quarter quarters, or "forties" (40 acres).

PROS AND CONS.   Thrower has examined the effects of different kinds of surveys on transportation lines and on administrative, property, and field units.[25] Although his sample areas were in Ohio, and he focused on the years 1875 and 1955, his conclusions appear to have wider applicability in the United States, and perhaps in other parts of the world as well. He selected two areas which were generally similar in their landforms, local relief, lithology, soils, climate, leading crops, and population density. Area S, in the northwestern part of the state, was surveyed systematically under the township and range land survey system, whereas area U, in the Virginia Military Survey area, was laid out unsystematically without any overall plan.

Nearly all of the roads in area S cut straight across country along section lines in a neat, rectangular grid, but the spiderweb road network of area U had virtually no relationship to survey lines. Area S had a denser road net than area U (plus a much greater number of larger and more expensive highway bridges), which is another way of saying that larger parts of area U were farther from roads, and farm driveways were longer. Fields and woodlots in area U had much greater variety in size and shape than the rectangular fields and woodlots of area S.

Many original survey lines were used as administrative (county,

24 William D. Pattison, *Beginnings of the American Rectangular Land Survey System, 1784–1800*, Research Paper No. 50 (Chicago: University of Chicago Department of Geography, 1957), has traced the idea of rectangular land survey back to the Roman scheme of centuriation, which is described in some detail in George Kish, "Centuratio: The Roman Rectangular Land Survey," *Surveying and Mapping*, Vol. 22 (1962), 233–44, but Norman J. W. Thrower, *Original Survey and Land Subdivision: A Comparative Study of the Form and Effect of Contrasting Cadastral Surveys*, Association of American Geographers Monograph No. 4 (Skokie, Ill.: Rand McNally, 1966), p. 8, said that early towns in the Indus valley had gridiron plans; they probably did not have much influence on Thomas Jefferson, because it is quite unlikely that he had ever heard of them.

25 Thrower, *Original Survey and Land Subdivision*, footnote 24.

township, or school district) boundaries and property lines in area S, but few were so used in area U. The property units in area U which were in two or more political jurisdictions were a real headache for taxation purposes (both paying and collecting). Furthermore, the property boundary lines in area U have been much more fluid than those in area S, and the Virginia Military District as a whole, although it covers only one-sixth of Ohio, has had more litigation over property boundaries than all the rest of the state combined.

Thrower has highlighted the strong points and the weaknesses of the township and range survey system. Criticism of this system has been terribly fashionable in recent years. It probably has necessitated a greater expenditure for such rural services as highways, telephone and power lines, mail delivery, and school bus routes. In rough, dissected country it has posed problems in the layout of transport lines, and it has encouraged field layout against the grain of the country, to the dismay of conservationists. Rural sociologists have argued that it has virtually ensured that the farm family would live in an isolated farmstead, which, depending upon your point of view, would either give them a bit of privacy from their neighbors, or would result in the "incomplete socialization of individuals" by denying them ready contact with members of other farm families.[26] Although the township and range survey system, like any other system of land survey, has its defects, T. Lynn Smith probably was guilty of hyperbole when he thundered, "From the standpoint of the social and economic welfare of the population on the land it is one of the most vicious modes ever devised for dividing lands."[27]

In fact, the township and range survey system is the best and simplest system of land division ever invented by the mind of mortal man (Fig. 1-1). It provides an excellent frame of reference for orientation, and it conveys a sense of neatness, order, and stability. It has obviated an enormous amount of litigation by facilitating brief but precise description of the exact location of any tract of land. It permitted alienation of land to individual owners in compact units of similar size, in keeping with the egalitarian doctrine of *isonomy*.[28]

It is patently ridiculous to assume that a rectangular survey system forces farmsteads to be isolated one from another. A square farm has four corners, and at each corner it is contiguous with at least two, and perhaps three, other farms. If farmers were genuinely concerned about their "incomplete socialization," they could have located their farmsteads at the corners of their properties, and farmsteads would be scattered across the countryside in clusters of three or four. The comparative rarity of such clusters indicates that other considerations are more important to farmers than the theories of rural sociologists.

[26] Alvin L. Bertrand, ed., *Rural Sociology: An Analysis of Contemporary Rural Life* (New York: McGraw-Hill, 1958), p. 162.
[27] T. Lynn Smith, *The Sociology of Rural Life*, 3rd ed. (New York: Harper and Row, 1953), p. 268.
[28] Land alienation in New England was based upon the aristocratic doctrine of *eunomy*, which allocated each man an acreage of land in accordance with his value or importance.

STRANGE SQUARES ON THE CHECKERBOARD. The theory of enforced isolation of farmsteads is closely related to the "checkerboard theory" of those who have assumed, apparently after only cursory examination of the idealized textbook diagram, that the township and range survey system results in "a checkerboard arrangement of fields and farms; properties of 160 acres, a quarter section, are divided into square 40-acre fields."[29] The actual situation is vastly more complex, as almost any aerial photograph of almost any area almost anywhere in the Middle West will clearly demonstrate.[30]

The square section (640 acres) and its square quarterings (160, 40, and 10 acres) are, to be sure, the basic blocks of land which comprise the fields and farms of township and range areas, but these basic blocks are combined into mosaics of remarkable variety. In 12 square miles in north central Indiana, for example, where I should have found 192 square 40-acre fields if the checkerboard theory were valid, I actually found only nine, and Thrower discovered an almost identical situation in the township and range survey area which he studied in Ohio.[31] Furthermore, Thrower found that only 9 percent of the properties in 1875 and 3 percent in 1955 consisted of 160 acres; the most common size, 80 acres, accounted for only about a quarter of all properties, and simple divisions of a section (160, 120, 80, 60, 40, and 20 acres) for little more than half.[32] Similarly, Kiefer found that a farm of 160 acres was the exception rather than the rule in Rush County, Indiana.[33]

THE STABILITY OF FARM BOUNDARIES. Not only are property units in the eastern Middle West smaller than many scholars seem to have realized, but their boundaries are remarkably stable. Thrower found a total of 649 miles of property boundaries in Ohio in 1875 and/or in 1955, of which 463 miles, or 70 percent of the total length, were the same in both years.[34] Kiefer's analysis of Rush County, Indiana, found a similar stability; many of the property lines in the modern plat book are the boundaries of the original units purchased from the General Land Office when the land was originally alienated in the 1820s and 30s.[35]

Although these are two isolated examples, they are the only instances I know of where an attempt has been made to examine the tenacity and permanency of property boundaries, and in both cases such boundary lines appear to have remained remarkably stable. We are forced to conclude, therefore, that the size of farms in the eastern Middle

[29] "Settlement Geography," Chap. 5 in Preston E. James and Clarence F. Jones, eds., *American Geography: Inventory and Prospect* (Syracuse, N.Y.: Syracuse University Press, 1954), p. 128.
[30] Marschner, *Land Use,* footnote 1, pp. 138–77.
[31] John Fraser Hart, "Field Patterns in Indiana," *Geographical Review,* Vol. 58 (1968), 463; and Thrower, *Original Survey and Land Subdivision,* footnote 24, p. 85.
[32] Thrower, *Original Survey and Land Subdivision,* footnote 24, pp. 55, 67, and 71.
[33] Wayne E. Kiefer, *Rush County, Indiana: A Study in Rural Settlement Geography,* Geographic Monograph Series, Vol. 2 (Bloomington: Indiana University Department of Geography, 1969), pp. 82–91.
[34] Thrower, *Original Survey and Land Subdivision,* footnote 24, p. 68.
[35] Kiefer, *Rush County, Indiana,* footnote 33, pp. 27–37 and 82–91.

West, at least, and presumably over a much larger area, has remained unchanged for at least a century, perhaps even longer, and that it was determined by the size of the unit of land purchased from the federal government when the land was first settled.

In order to understand the present size of farms, therefore, it would seem appropriate to inquire when the land was originally settled, and to examine the changing public land alienation policies of the federal government in relation to the westward spread of settlement across the United States. Before 1862, when the Homestead Act was passed, the prospective settler had to purchase the land, rather than receiving it free. Presumably the majority of settlers were short of ready cash and bought only the minimal acreage required by law, so policy concerning the minimal size of unit which could be purchased is of special importance.

The Alienation of Public Lands

Although historians have carefully examined the public land policies of the United States, it is rather surprising that the alienation of the public lands has received such slight attention in the literature of American geography.[36] The records of the General Land Office are remarkably detailed, but the intensive geographical analysis which they deserve has hardly begun.[37] We need far more detailed and intensive investigations of the interaction of land alienation policies and initial settlement, and their impact upon the contemporary rural landscape of the United States.[38]

In 1789 the new United States was a pitifully poor nation, and the western lands were one of its principal assets. The early land policies of the young republic were designed to obtain revenue from the sale of these lands, rather than to encourage their settlement, but for decades the halls of Congress rang with debates over the minimal price at which land should be sold, and the minimal acreage which a buyer should be required to purchase.

In the East it was argued that revenue could be obtained most efficiently by selling the land in large blocks. The cost of survey would be increased appreciably if the land had to be divided into small parcels, and the sale price would be reduced if parcels were so small that wealthy buyers could not be troubled to bid on them. Furthermore, it would be more difficult for prospective purchasers to select only the better land

---

[36] A number of studies of public land policies by historians are cited in John Fraser Hart, "The Westward Movement of the Frontier, 1820–1860," in *Man and Cultural Heritage,* H. J. Walker and W. G. Haag, eds., Geoscience and Man, Vol. 5 (Baton Rouge: Louisiana State University School of Geoscience, 1974).

[37] Kiefer, *Rush County, Indiana,* footnote 33, and Charles E. Dingman, "Land Alienation in Houston County, Minnesota: Preferences in Land Selection," *Geographical Bulletin,* Vol. 4 (1972), 45–49, provide examples of the use that can be made of General Land Office records of land alienation.

[38] C. Barron McIntosh, "Forest Lieu Selections in the Nebraska Sand Hills," *Annals,* Association of American Geographers, Vol. 64 (1974), 87–99, illustrates the way in which land alienation policies can aid or thwart the development of certain types of agricultural operations.

and leave the poorer if the land were sold in large blocks, and the danger of Indian attack would be lessened if settlements were grouped on blocks of land rather than scattered through the wilderness. There was also some concern in the older states about the loss of labor if cheap land in small blocks at low prices stimulated migration to the West.

The settlers in the West, however, believed that they were performing a patriotic duty when they tamed the wilderness, and they wanted to have the land free, or at the very least, in tracts so small and at prices so low that every man could afford the "threshold price" (the minimal acreage of land multiplied by the minimal price) of a farm. "Hence," said Hibbard, "the long list of acts condoning the violation of the newly enacted land laws."[39] The minimal amount of land offered for sale was gradually whittled down from whole townships and whole sections of 640 acres in alternating townships, which had been stipulated in the Land Ordinance of 1785, to 320-acre half-sections in 1800, 160-acre quarter-sections in 1804, 80-acre tracts in 1820, and 40-acre tracts in 1832.

In the early years government land sales had to compete with those of the states, which had land of their own to sell, and with holders of military warrants, most of whom were interested only in converting their land warrants into cash as quickly as possible. By 1820, however, these sources of cheap land had more or less dried up, and it was partly for this reason that the minimal purchase for federal land was reduced from $2.00 an acre, where it had been set in 1796, to $1.25 an acre, which meant that the "threshold price" for a farm was reduced from $1,280 (640 acres at $2 an acre) in 1796, to $100 (80 acres at $1.25 an acre) in 1820, and to $50 (40 acres at $1.25 an acre) in 1832, where it remained until the Homestead Act was passed in 1862.

GROSS GEOGRAPHICAL PATTERNS OF LAND ALIENATION.   Approximately half of the federal land in the state of Ohio had been alienated before 1820, while the minimal purchase unit was still 160 acres. Most of the remainder in Ohio, and at least half of the federal land in Indiana, Illinois, Michigan, and Mississippi, was sold during the land boom of the 1830s, when the minimal unit of purchase was only 40 acres (Fig. 4–2). In Mississippi much of the land was purchased in large blocks by speculators, but it would appear reasonable to assume that a considerable portion of Indiana, Illinois, and southern Michigan was alienated in small blocks of 40 or 80 acres.[40]

The complexity of the actual pattern of size of purchase unit is indicated by Kiefer's study of Rush County, Indiana.[41] The county was opened to settlement shortly after the minimal mandatory purchase unit was reduced from 160 acres to 80 acres in 1820, and most of the land was purchased in 80-acre tracts. The majority of the purchase tracts larger than 80 acres are in the southeastern part of the county, on the

39 Benjamin Horace Hibbard, *A History of the Public Land Policies* (New York: Macmillan, 1924), p. 5.
40 Roy M. Robbins, *Our Landed Heritage: The Public Domain, 1776–1936* (Princeton, N.J.: Princeton University Press, 1942), p. 61.
41 Kiefer, *Rush County, Indiana,* footnote 33, pp. 27–37.

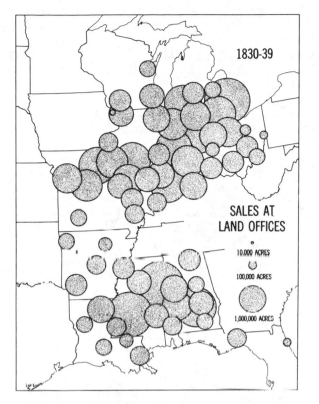

FIG. 4–2. *Land sales at land offices in the United States, 1830–1839. Reproduced by permission from H. J. Walker and W. G. Haag, eds.,* Man and Cultural Heritage, Geoscience and Man, *Vol. 5 (Baton Rouge: Louisiana University School of Geoscience, 1974).*

better land which was purchased in the early 1820s (Fig. 1–2, p. 8). By the time the minimal size of purchase unit had been reduced to 40 acres, in 1832, the only unclaimed land remaining was a few odd bits and pieces in the northern and western parts of the county, but a surprisingly large amount of this land was purchased in 40-acre tracts. Conversely, virtually all of the land which was purchased in tracts of 40 acres was land which had not been sold when the law was changed in 1832.

The Homestead Act of 1862, which gave 160 acres free to any man who would live on the land for five years, was almost completely irrelevant in an arc of states running from Michigan through Wisconsin, Iowa, Missouri, and Mississippi into Albama, because most of the remaining federal land in these states had been purchased in the land boom of the 1850s (Fig. 4–3). Apart from Florida, most of the federal land east of the Mississippi River had already been alienated by the time the Homestead Act was passed, and even in the first tier of states

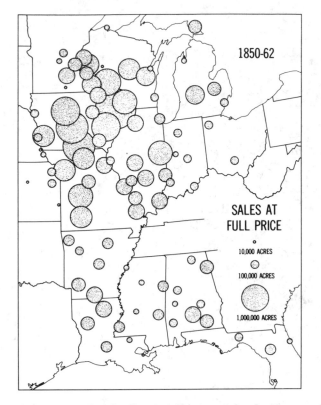

FIG. 4–3.   *Land sales at full price at land offices in the United States, 1850–1862; sales at graduated prices are not shown. Reproduced by permission from H. J. Walker and W. G. Haag, eds.,* Man and Cultural Heritage, Geoscience and Man, *Vol. 5 (Baton Rouge: Louisiana State University School of Geoscience, 1974).*

west of the river only Minnesota, Arkansas, and Louisiana had sizable areas available for homesteading.[42] The 160-acre farms provided under the Homestead Act must be sought in the Great Plains and Rocky Mountain states, because most of the Corn Belt and adjacent areas had already been settled before the act was passed.

THE SPREAD OF POPULATION.   The patterns shown on the maps of land alienation are repeated, in essence, on maps showing the spread of settlement, as indicated by the density of population at successive decennial censuses. It has been customary to define the frontier as the outer limit of the area with a population density of at least two persons per square mile.[43] If it is reasonable to assume, however, that a pioneer family consisted of a man, his wife, and at least two children, then a

[42] John Fraser Hart, "The Middle West," *Annals,* Association of American Geographers, Vol. 62 (1972), 263.
[43] This definition has been challenged in Hart, "The Westward Movement of the Frontier," footnote 36.

density of two persons per square mile would represent only one family (and presumably only one home and only one farm) on each two square miles of land, which would provide them with a fair amount of elbow room, but would be pretty sparse settlement.

Perhaps a density of two persons per square mile might properly represent the outer limit of settlement, but one might argue that an area had not actually been *occupied* until it had six persons (or a family and a half) per square mile, and that it had not really been *settled* until it had 18 persons (or four and a half families) per square mile. The use of these definitions, admittedly, would suggest that fairly large areas in the West have never truly been occupied and settled, a suggestion which is not completely unreasonable, however much it might affront local pride. In the great majority of essentially rural minor civil divisions in the eastern half of the nation these definitions accord fairly well with the normal model of population growth: a rapid initial spurt, from 2 to 6 to 18 or more persons per square mile in the first few decades, and then stability or a very gentle decline.

The occupied area of the United States in 1790 extended along the Eastern Seaboard from central New England to North Carolina, with a major lobe extending southwestward along the Piedmont into Georgia (Fig. 4-4). Outpost areas around Pittsburgh and in northeastern Tennessee lay beyond the interior margin, which ran along the foothills of the Folded Appalachians and the Blue Ridge Mountains; the South Carolina coast near Charleston was another isolated outpost.

Expansion from this original nucleus was spasmodic, and "negative areas" in northern New England, the Adirondacks, Appalachia, and the southern Coastal Plain continued to repel occupation as late as 1870. In the northeast expansion into the uplands was slow but steady, and a bit more rapid along the Champlain and Mohawk lowlands. In the west the principal thrust was along the Ohio Valley toward early settlements in the Bluegrass country and the Nashville Basin, and up the major tributaries to the north, with outposts along the Mississippi and Missouri Rivers.

By 1820 most of Ohio had been occupied, and the next two decades witnessed fairly rapid expansion overland from the major rivers of the Middle West. Two large poorly drained areas, the Black Swamp of northwestern Ohio and the Grand Prairie of northeastern Illinois, were bypassed briefly, but by 1860 the outer limit of occupance, which was not greatly extended during the Civil War, described a ragged arc running from southern Michigan and Wisconsin through southeastern Minnesota and central Iowa into the eastern margins of Nebraska and Kansas. Occupation of the land farther south had extended westward to the Mississippi River fairly rapidly in the 1820s and 30s, but large parts of Arkansas, Louisiana, and eastern Texas were not occupied until just before the outbreak of the Civil War, and the Ozarks remained a negative area as late as 1870.

The patterns on the map of the extension of settled areas are similar to, but even more complex than, those on the map of occupied areas

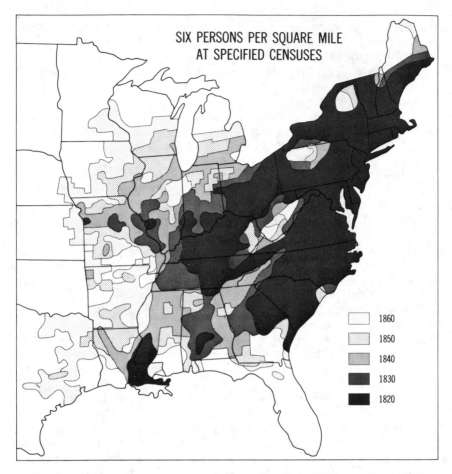

SIX PERSONS PER SQUARE MILE
AT SPECIFIED CENSUSES

1860
1850
1840
1830
1820

FIG. 4–4. *The spread of occupance in the United States, 1820–1860, as measured by a density of at least six persons per square mile. Reproduced by permission from H. J. Walker and W. G. Haag, eds.,* Man and Cultural Heritage, Geoscience and Man, Vol. 5 *(Baton Rouge: Louisiana State University School of Geoscience, 1974).*

(Fig. 4–5). The negative areas are larger, and remained unsettled longer. Settled areas have developed fairly rapidly in outlying "pockets," or growth nodes, on the better lands, but these have become connected with the main areas of settlement along strips which have grown only slowly. The first strip of continuously settled area connecting the main settled areas on the East Coast with those in the Middle West followed the Erie Canal routeway, and the more direct connection across the Appalachians was not developed until 1850. Settled areas in the South expanded fitfully along the Piedmont, and the link-up with early coastal settlements around Charleston and Savannah came fairly late. By 1850 a continuous belt of settled areas extended along the Great Valley, and

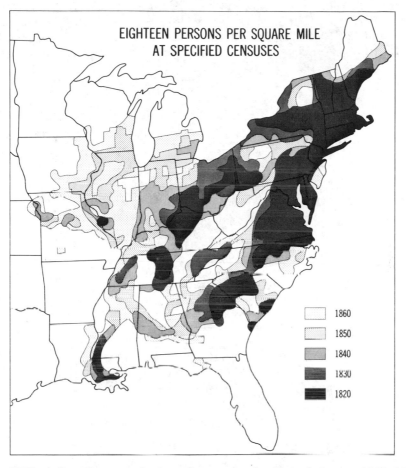

EIGHTEEN PERSONS PER SQUARE MILE
AT SPECIFIED CENSUSES

	1860
	1850
	1840
	1830
	1820

FIG. 4–5. The spread of settlement in the United States, 1820–1860, as measured by a density of at least 18 persons per square mile. Reproduced by permission from H. J. Walker and W. G. Haag, eds., Man and Cultural Heritage, Geoscience and Man, *Vol. 5 (Baton Rouge: Louisiana State University School of Geoscience, 1974).*

the settled areas of the South had been connected with those in the Middle West around the southwestern end of Appalachia, but the settled areas of the lower Mississippi valley remained isolated.

The census of 1870 was taken eight years after the passage of the Homestead Act, and only five years after the end of the Civil War had permitted settlers to begin to take advantage of it on a grand scale. By 1870 the eastern half of Iowa, northern Missouri, and the southern parts of Wisconsin and Michigan had already been settled. Most of Minnesota, the Dakotas, Nebraska, and Kansas lay open to the settler, as did less attractive areas in the boreal forests of the north, in the hills, swamps, and sandy lands of the south, and in areas farther east which had long since been bypassed, but most of the Middle West had already been occupied and settled before the Homestead Act had had time to become really effective.

C H A P T E R 5 *farm size*

and farm tenure

Any rural landscape is the product of a host of independent deci-sions made by the multitudes of individuals (including groups behaving as though they were individuals) who control and have controlled the individual pieces of land. If we wish to understand the look of the land, we must identify the individuals who have made the decisions, and we must try to learn why they made them as they did. In order to do so, we need four basic kinds of information: (1) how the land is divided into individual units (which, in most rural areas, means farms); (2) how large these units are, and how they have been subdivided; (3) who owns and who operates them; and (4) what factors influence the decisions which the owner and operator make concerning them. We looked at systems of land division in Chapters 3 and 4. In this chapter we will consider the size and subdivision of farms, and systems of land tenure; the next will look at farm enlargement and farm management systems, and Chapter 7 will deal with the factors influencing the decisions which have such a profound effect upon the look of the land.

The Size of Farms

Official statistics on the average size of farm in the United States are virtually meaningless. Every statistician knows that averages are dis-torted by values for extreme cases, and the census definition of a farm includes plenty of extreme cases at both ends of the spectrum. Most of our largest "farms," for example, are livestock ranches on some of the poorest land in the nation; if it had been any poorer this land probably would have been made into reservations and given back to the Indians. Ranchers on such land customarily measure their holdings in sections (square miles) rather than acres. These holdings are so large, and their numbers are so few, that fluctuations in their average size mean almost nothing, yet such fluctuations can significantly influence the national average.[1]

The statistical aberrations caused by the very large farms, which are mainly in the West on the drier lands, are much easier to cope with than those posed by "mini-farms" in all parts of the nation. At least a million of these mini-farms can be considered farms only by courtesy of the completely unrealistic definition of a farm which is used by the Bureau of the Census. The specific details of the official definition have varied a bit from census to census, but in essence any piece of land has been classified as a farm for census purposes if agricultural operations were (or could have been) conducted upon it, and if it met remarkably low minimal criteria of size (three to ten acres) and value of agricultural products ($50 to $250) produced on or sold from it.

These low minima have produced greatly inflated estimates of the number of farms in the United States, with a corresponding reduction in the estimated average size of farm. In 1959, for example, the minimal size required before a place could be included in the census as a farm was increased from three to ten acres. This change in definition excluded 232,059 farms, which contained 5,776,945 acres of land; if they had been included they would have accounted for 6 percent of the nation's farms (or more than one farm of every 20), but only .05 percent of its farmland (or about one of every 200 acres). Between 1954 and 1959 the average size of farm in the United States officially increased from 242 acres to 303 acres, but more than a quarter of this increase (17 of 61 acres) was entirely a result of the change in definition, which excluded almost 250,000 places that should never have been classified as farms to begin with.

Despite the change in definition, the smallest size categories still contain a very large proportion of the nation's farms and a very small part of its farmland, whereas just the reverse is true of the largest categories. In 1964 farms of less than 50 acres accounted for about a quarter of the nation's farms but less than 2 percent of the farmland. Farms of 500 or more acres, conversely, accounted for only 10 percent of the farms, but almost two-thirds of the farmland. The same general pattern emerges when farms are classified in terms of income. In 1964 more than 1.3 million farms sold less than $2,500 worth of farm products. These farms produced less than 4 percent of the national total on 17 percent of the farmland, but they accounted for more than 40 percent of all farms. At the other extreme, less than a quarter of all farms sold more than $10,000 worth of farm products, but these farms contained almost two-thirds of

---

[1] Esmeralda County, Nevada, led the nation in 1964 with an average farm size of 88,487.3 acres; four of these farms would have been as large as an Iowa county, and eight of them would have blanketed the entire state of Rhode Island. In 1959 the average size of farm in Esmeralda County had been 115,928.6 acres; the loss of 27,441.3 acres in average farm size might be attributed to the fact that the number of farms in the county increased from 18 in 1959 to 23 in 1964. In Brewster County, Texas, which had 106 farms in both years, the average size increased from 21,914.1 acres in 1959 to 24,536.8 in 1964, an increase of 2,622.7 acres per farm.

all farmland and produced more than four-fifths of all farm products sold. A rule of thumb within the U.S. Department of Agriculture during the 1960s held that a farmer needed an annual net income of $2,500 to maintain a minimum decent level of living, and that he probably would have to sell at least $10,000 worth of farm products to achieve such a net.[2] In short, the United States actually had only 870,000 "real" farms in 1964, rather than the 3,158,000 reported in the census.

How can one explain the use of such an unrealistic definition of a farm for census purposes? Cynical souls have suggested that perhaps the Department of Agriculture has hoped to justify its continued existence by attempting to conceal the magnitude of the decline in number of farms over the last few decades; certainly the number of employees of the Department of Agriculture has not declined proportionately with the decline in the number of farms (Fig. 5-1). A more charitable explanation

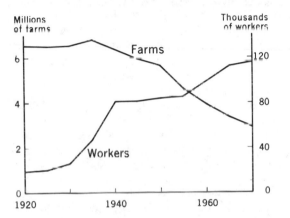

FIG. 5–1. *Millions of farms in the United States, 1920–1970, and thousands of paid civilian employees of the U. S. Department of Agriculture. Compiled from data in the* Statistical Abstract of the United States, *various years.*

[2] Calvin L. Beale, "The Negro in American Agriculture," in John P. Davis, ed., *The American Negro Reference Book* (Englewood Cliffs, N.J.: Prentice-Hall, 1966), p. 179. At this point, in order to prevent any misunderstanding, I should like to pay explicit, heartfelt tribute to the superb job which the U.S. Bureau of the Census has done in collecting and publishing statistical data on the fourth largest country in the world, data whose richness and reliability are the envy of scholars in every other nation. The Bureau has the task of collecting information on some 200 million people, 60 million housing units, 12 million business establishments, 3 million farms, 300 thousand manufacturing companies, 90 thousand units of government, and 40 thousand mining companies, not to mention fisheries, and transportation on water, land, and in the air. It performs this task with truly remarkable speed and accuracy to provide an unrivaled source of information for any student of the United States. It cannot hope to satisfy the whims and foibles of every user, in large part because of the continental scale at which it operates; it must use definitions which are equally applicable in the swamps of Florida, the forests of Washington, the potato fields of Maine, the citrus groves of California, and all places in between. Anyone who uses census data must understand how they were collected, and the definitions which were used in collecting them, but for his efforts he will be rewarded with the world's richest bonanza of statistical information.

would suggest that human beings live on each of these farms, no matter how uneconomic they may be, that most of these people derive some part of their living from the land, and that they have some impact on the way the land is used and how it looks. One might also argue that it is better to collect too much information, rather than too little; the student can always ignore data which are not relevant to his purpose, but he can never hope to use data which have not been collected.

An Index of Farm Size

How can the vast storehouse of census data be tapped to provide better insights into geographical and historical variations in the size of farms? Some very significant indices of farm size probably could be devised by using data on the value of farm products sold, but I have not used these data for two reasons: first, comparable data are available only for the period since World War II, and second, it is extremely difficult to correlate these data geographically with the areal extent of farm units and the look of the land. A further reason is the fact that their utility and applicability have not, to the best of my knowledge, been explored by geographers, although I suspect that their careful exploration would reveal a fruitful field for geographic research.

In the humid East (where data on farm acreage are not confounded by extremely large holdings of range land which, at best, has only limited agricultural value) a useful index of farm size is the percentage of farmland which is in farms larger than a specified acreage. Such an index must be calculated on a cumulative ("more than" or "less than") basis, because individual size categories may retain a fairly stable percentage of farmland even though farms are growing in size.[3] In the United States as a whole, for example, the proportion of land in farms of 180 to 499 acres remained fairly close to a quarter of all farmland between 1900 and 1964, because smaller farms were being enlarged and moving into this category as larger ones were acquiring more acreage and moving out of it.

The use of such an index may be illustrated by data from eight states in the Middle West, where the size of farms did not change very much before World War II (Fig. 5–2). The total acreage of farmland remained slightly more than 200 million acres from 1900 to 1945, but has been declining steadily ever since. Before 1945 the proportion of farmland in the larger size categories was increasing, but not very rapidly, and the proportion in the smaller size categories was dropping. Since 1945 these trends have accelerated. and the size of farms has been increasing rapidly; in 1940 more than half the farmland in the Middle West was in farms smaller than 180 acres, but by 1964 more than half was in farms larger than 260 acres. The curves for the various size categories converge on 1984; by that year, if present trends continue, the total farmland area

---

[3] Everett G. Smith, Jr., "Road Functions in a Changing Rural Environment," unpublished doctoral dissertation, University of Minnesota, 1962, pp. 81–86.

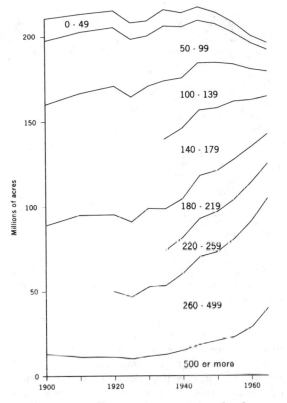

FIG. 5–2. *Millions of acres of farmland in specified size categories in eight states (Illinois, Indiana, Iowa, Michigan, Minnesota, Missouri, Ohio, and Wisconsin) of the Middle West, 1900–1964. Reproduced by permission of the publisher from John Fraser Hart, ed., Regions of the United States (New York: Harper & Row, 1972), 273.*

of the Middle West will have dropped to 170 million acres, and no farm will be smaller than 220 acres.

The geographical pattern of farm size has remained relatively stable; in fact, a remarkably accurate "prediction" of the size of farms in each Indiana county could have been based upon knowledge of farm size in 1945 (Fig. 5–3). The percentage of each county's land which was in farms of 180 acres or more in 1945 had a coefficient of correlation of +0.92 with the percentage which was in farms of 260 acres or more in 1964.

The Operating Unit

The man who owns a farm is not necessarily the man who operates it. The owner is the person who holds legal title to the land, but the operator is the person who does all the work or who directly supervises it. On any given farm the functions of owner and operator may be com-

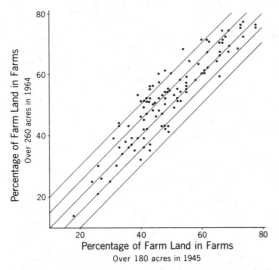

FIG. 5–3. The percentage of farmland in farms of 260 acres or larger in Indiana counties in 1964 had a coefficient of correlation of +0.92 with the percentage of land in farms of 180 acres or more in 1945.

bined in the same person, although they need not be. The man who owns the land may operate it himself, as owner-operator; he may entrust its operation to a manager, overseer, or bailiff; or he may lease some or all of it to another person as tenant. The man who operates the farm may be full owner of all the land he operates; he may be part owner who owns part of his land and rents the remainder from someone else; he may be a tenant who rents all of his land; or he may be a hired manager. In the 1964 Census of Agriculture approximately 58 percent of all farm operators in the United States were classified as full owners, 25 percent as part owners, 17 percent as tenants, and less than 1 percent as managers.

If the functions of ownership and operation are not combined in the same person, the student of the rural landscape must be aware of the fundamental distinction between ownership units and operating units. As a general rule, the owner of a farm is responsible for the provision and maintenance of its permanent structures, such as barns and fences, and thus he has an important impact on the look of the land even though he is not actually involved in working on it. If he leases a farm he may stipulate the kinds of livestock which may be kept on it, the kinds of crops which may be produced, and perhaps even the rotation in which they must be grown. In short, he may well decide how the land will be used, as well as how it will look.

The boundaries of ownership units tend to be a bit more permanent than those of operating units, because they are defined by legal titles in the public records, and thus can be changed only at the expense of services from a lawyer, title abstractor, and other functionaries. These boundaries, therefore, are somewhat easier to identify and map, although the compilation of an ownership map from legal records is a chore whose tediousness must be experienced to be truly appreciated. The task is made less onerous in much of the Middle West by the existence of county atlases and plat books, sources of information whose riches have scarcely been tapped by geographers.[4]

It is completely incorrect, however, to assume that a map of ownership units is also a map of operating units, because any landowner may have rented part or all of his land to another farm operator. The boundaries of operating units are even more difficult to identify and map than those of ownership units, because a tenancy arrangement may be defined by a contract, a lease, or only an informal oral agreement and a handshake; it expires automatically after a stipulated period of time, and then is subject to renegotiation and change. Quite commonly the boundaries of operating units can be determined only by interviewing each and every person who owns and/or leases in an area.[5]

Despite the difficulty of identifying and mapping them, operating units are critically important to anyone who wishes to understand the rural landscape, because the operator makes many, if not most, of the decisions which determine how the land will be used and how it will look. The operator decides what crops will be grown (and in what rotation), what livestock will be kept, and what type of farming will be practiced; this fact is recognized by most censuses of agriculture, which collect and publish data on operating units rather than ownership units. Although a landlord might wish to exercise some control over the decisions of his tenants, he is seldom in a position to do so very effectively, because the tenant who is offered a lease he considers too restrictive is perfectly free to shop around for one he likes better. Conversely, a tenant often takes the initiative in urging his landlord to invest in new permanent structures for the farm, and thus he plays an additional role in influencing the look of the land.

FIELD PATTERNS. One of the most striking manifestations of farm operating units upon the landscape, whether as seen from the air or viewed on aerial photographs, is in their subdivision into fields. Although field patterns in most areas are remarkably complex, each and every field has its present size and shape because at one time that particular size and shape made some sense to somebody. As a general rule, livestock farmers prefer fields which are small and squarish, because these permit better control of grazing and breeding, whereas crop farmers are more inclined toward long and rather narrow fields, which are better suited to the use of machinery. As an even more general rule, large farms tend to be

---

[4] Norman J. W. Thrower, "The County Atlas of the United States," *Surveying and Mapping*, Vol. 21 (1961), 365–73.

[5] Smith, "Road Functions in a Changing Rural Environment," footnote 3, p. 79. Sometimes even these "informants" are not quite as well informed as one might wish; in densely settled England, for example, I learned of an area of roughly 800 acres between two upland farms which was neither claimed nor used by the farmer on either side; John Fraser Hart, *The British Moorlands: A Problem in Land Utilization* (Athens: University of Georgia Press, 1955), p. 71.

divided into large fields, and small farms into small ones, because there is an upper limit (often related to the crop rotation practiced) to the number of fields which one operator can manage efficiently. Very few farms, no matter how large they are, have more than eight or ten fields, and few have less than three. In the absence of any real information on the subject, one might hazard a guess that the great majority of farms in the United States are probably subdivided into four or five fields.

In most areas the field pattern is closely related to the system under which the land was surveyed. Within a given system of land survey, variations in the form of the land surface are probably the primary determinant of variations in field patterns. A rather striking example of the influence of land forms is provided by fields in which the crops are planted in rows or strips along the contour, at right angles to the natural slope of the land.

Field patterns may also be influenced by the farming system. In dry farming areas of the West, for example, alternating strips of land are cultivated and left fallow, which gives the countryside a distinctive striped appearance. In Indiana an area of corn-hog farming had small square fields, an area of diversified small farms had small rectangular fields, and an area of highly mechanized cash-grain farming had fields which were appreciably larger and more elongated than fields in the other two areas.[6] In the cash-grain farming area fences were being removed and fields enlarged so that the rows in each field could be as long as possible, and the farmer would waste a minimal amount of time turning his machinery at the end of the field.

This modification of fields to fit machines is unusual, because the layout of fields, as a general rule, is even more conservative, and changes even more slowly, than agricultural practices. The field pattern of an area often reflects past agricultural practices better than it reflects those of today. As a specific example, most of the "envelope" fields of the Middle West received their distinctive pattern when they were laid out for horse plowing, to avoid having to lift the plow out of the ground more often than necessary, but the same patterns are still being followed by tractor plows.[7] Plowing starts at the outside of the field, and works concentrically inward, but the imperfectly plowed diagonal strips where the plow was turned at the corners must be replowed after the rest of the work has been completed.

In Europe, where geographers have studied them much more intensively, field patterns appear to be even more persistent than they are in the United States. Fields in the United States are merely subdivisions of farm operating units, but Houston seems to have assumed that field boundaries in most parts of Europe are more permanent than farm boundaries, because most of his chapter on "The Rural Landscape" is an

---

[6] John Fraser Hart, "Field Patterns in Indiana," *Geographical Review*, Vol. 58 (1968), 450–71.

[7] I. F. Reed, *Laying Out Fields for Tractor Plowing*, Farmers' Bulletin No. 1054 (Washington, D.C.: U.S. Department of Agriculture, 1954).

extended discussion of field patterns.[8] The idea that field patterns have remained stable for a century or more, but operating units have changed repeatedly as individual fields have been added or lost, is supported by work in the Chilterns, and by Williams' study of a large parish in south-western England.[9] Field patterns have been stabilized in many parts of Europe because the individual fields are enclosed by such difficult-to-remove boundaries as hedgerows, earthen banks, or stone walls.

The size and shape of fields and farms varies enormously in Europe, even within short distances. Houston said that variations in field patterns were the product of two processes, a slow and gradual "evolution" through time, and a sudden and dramatic "creation," such as the systematic changes associated with the enclosure movement.[10] Coppock suggested that fields laid out after enclosure have straighter boundaries and more regular shapes than earlier fields, but he also noted a tendency toward elongated farms running across the grain of the country in areas such as the scarplands where different kinds of terrain and soil are in close proximity.[11] The success of a farm often depends on such a layout, which includes several different kinds of land within its limits; Stamp concluded his detailed, and abundantly illustrated, description of farm and field patterns in Britain with the suggestion that an optimum farm operating unit under British conditions in 1948 would consist of 12 ten-acre fields.[12]

CIRCULATION FACILITIES. Every farm operating unit needs circulation facilities to keep the farmer and his family in contact with the rest of the world. Telephone and electricity (for the radio and television set) lines permit the exchange of ideas. Paths, trails, lanes, roads, and highways provide for the movement of people and the transportation of goods. The examination of roads can contribute to an understanding of the countryside in two ways. The impact of their physical character (construction, windings, gradients, even the color of their surface materials) on the landscape is interesting, though of relatively minor importance, but the modern highway may be quite demanding of land; the old 16.5 foot right-of-way of horse and buggy days required only two acres of land per mile of road, but the 300-foot minimum right-of-way stipulated by the interstate highway system in the United States occupies 36.4 acres of land for each mile of highway.[13]

The function of rural roads, as indicated by their arrangement,

[8] J. M. Houston, *A Social Geography of Europe* (London: Duckworth, 1953), pp. 49–79.

[9] J. T. Coppock, "Farms and Fields in the Chilterns," *Erdkunde*, Vol. 14 (1960), 134–46; and W. M. Williams, "The Social Study of Family Farming," *Geographical Journal*, Vol. 129 (1963), 63–75.

[10] Houston, *A Social Geography of Europe*, footnote 8, p. 49.

[11] Coppock, "Farms and Fields in the Chilterns," footnote 9, pp. 140 and 143.

[12] L. Dudley Stamp, *The Land of Britain: Its Use and Misuse* (London: Longmans, Green, 1948), pp. 335 and 350.

[13] Brunhes has gone on at great length about the physical appearance of roads; Jean Brunhes, *Human Geography*, abridged edition by Mme. M. Jean-Brunhes Delamarre and Pierre Deffontaines, trans. by Ernest F. Row (Skokie, Ill.: Rand McNally, 1952), pp. 55–58.

density, and use, is probably more important than their appearance. During the 1920s in Illinois, for example, "road turnings" were a clue to the boundaries of the trade areas of small central places.[14] The rural road system of the United States had been fairly well set by 1920, and few roads have been added or removed since then.[15] Most of the rural roads follow section lines in those parts of the Middle West which were laid out according to the township and range system of land survey, and this has produced a density of roughly two miles of rural road per square mile of land surface. To an ever increasing degree, these "farm to market" roads serve rural nonfarm people who have chosen to live in the country and commute to jobs 10, 20, or even 75 miles away.

The Family Farm and Farm Tenancy

Although the proportion of "farms" in the United States which are owner-operated has remained fairly stable between 50 and 60 percent since 1900, the percentage of farmland controlled by owner-operators has been declining fairly steadily since the First World War, and an increasing percentage is in the hands of part owners. Approximately half of the nation's farmland was operated by its owners in 1920, but by 1964 this figure had dropped to just over a quarter, whereas the proportion of farmland operated by part owners had increased from one-seventh to almost one-half. The proportion of farmland operated by tenants has dropped from around a quarter to just over 10 percent since the Second World War, but when tenants are combined with part owners it is evident that almost two-thirds of the farmland in the United States in 1964 was controlled by men who were tenants on at least part of the land they were farming.

The notion of farm tenancy is disturbing, or even shocking, to many Americans. It is the very antithesis of the Family Farm, an ideal which has taken its place close to God, Motherhood, and the Star-Spangled Banner in the national pantheon. The ideology of the family farm is an amalgam of many of the traditional frontier values: the self-made man is better than the well-born one, so every man should have enough land to enable him to prove himself; he should own his land in fee simple, and unencumbered by debts; his farm should be large enough to provide him with a decent livelihood, so that he can accept responsibility for the economic security of himself and his family; he and his family should provide a substantial share of the labor on it, but they should receive a fair return for their efforts (the prices farmers get should bear a reasonable relationship to the prices farmer have to pay); their income should be related to their diligence, because hard work is a virtue; and the worth of a man may be measured by the amount of money he makes.

Generations of Fourth of July orators have rumbled on about the virtues of individual farm ownership as one of the cornerstones of a free

14 Stanley D. Dodge, "Bureau and the Princeton Community," *Annals*, Association of American Geographers, Vol. 22 (1932), 159–200.
15 Smith, "Road Functions in a Changing Rural Environment," footnote 3.

and democratic society, and an enormous amount of sentimental drivel has been written about family farms, much of it quite obviously by those who haven't the faintest idea as to what a family farm really is, or how rapidly they have been changing. A family farm is simply an operating unit which provides an adequate level of living in return for the labor of a father and son, with a hired hand at certain stages of the demographic cycle (when the son is too young to be of much help, for example, or when the father is too old).

Neither size nor ownership are stipulated in this definition of a family farm, nor should they be. The amount of land that will provide a reasonable farm income under intensive cultivation in an irrigated area in the West would hardly be enough to keep a single cow alive on adjacent sagebrush-covered range land. Livestock ranches and highly mechanized grain farms both require and can use more land than farms which specialize in labor-intensive vegetables, fruit, tobacco, poultry, and dairy products, which demand individual attention and considerable hand labor. Recent changes in agricultural technology not only enable the individual worker to care for a greater amount of land, but also require him to do so if he is to receive an adequate return for his efforts. No matter what the farming system, the amount of land which was large enough for a family farm a generation ago has become, or is becoming, too small, and the size of the operating unit must be enlarged.

The ideal way to enlarge a farm operation is to buy more land, but many a farmer is in no position to do so. Like most of the rest of us, a farmer seldom has as much money as he would like to have, or even as much as he needs, and the only way he can get more land is to rent it. But he may have to rent land even if he does have enough money to buy it, because none of his neighbors is willing to sell. Some of them may think of their land as an investment which is too valuable to part with, or they may simply enjoy the prestige of landownership. Others may have a sentimental attachment to the old home place, which has been in the family such a long time.[16] The farmer himself can understand these attitudes, but cold-blooded city economists are often baffled, irritated, and impatient when they discover that many rural landowners have a strong, deep-down-in-the-bones, feeling that it is somehow wrong, even "sinful," to part with a piece of land for mere money; perhaps economists are the ones who are irrational.

The farmer who rents the land he needs to expand his operation, in order to keep his family farm from degenerating into an undersized unit, often discovers that tenancy is not really so bad, after all. He can put his capital to work earning money for him, rather than having it tied up in the land, and this can be pretty important for the young farmer or for the one with limited resources.[17] The size to which he can enlarge his

16 Howard F. Gregor, "A Sample Study of the California Ranch," *Annals,* Association of American Geographers, Vol. 41 (1951), 292.
17 The successful modern farmer is one who is in debt, but he must be a skillful money manager; C. Edward Harshbarger, "The Role of Financial Management in Agriculture," *Monthly Review, Federal Reserve Bank of Kansas City,* July-August 1972, pp. 14–20.

operation is limited only by the amount of land he can find to rent, and his ability to farm it, rather than by the amount of capital he has or can borrow. And his neighbors, instead of scorning him because he is a tenant, may envy him the size of his operation, which enables him to specialize and thus to achieve the economies of large-scale mechanized production.

Tenancy rates tend to be highest in areas of cash crop production, where a quick return can be gotten from the land, but both land and equipment are costly. A high rate of tenancy, for example, has been traditional in cash-grain (corn and soybean) farming areas in central Illinois, in wheat farming areas on the Great Plains, and in cotton and tobacco farming areas in the South.

New forms of tenancy may be evolving on large field-crop farms in some irrigated areas of the West. In specialized vegetable-producing districts such as the Imperial Valley of southern California, for example, the field-crop farmer practices a regular crop rotation, with a certain portion of his land in vegetables each year, but he does not wish to have the bother of growing vegetables himself. He rents his "vegetable land" (a different field each year) to a "vegetable man," who provides the necessary machinery, hires the labor, markets the crop, and takes all the other risks of vegetable production. This system assures the farmer of a regular income and provides the vegetable man, in return for the risks he takes, with land and the chance to make a killing or to lose his shirt in the volatile vegetable market. Some farmers in the Coachella Valley even rent their alfalfa land to dairy "farmers," who own no land whatsoever, but simply move their herds from one rented pasture to another.[18]

Tenancy, admittedly, does not work equally well in all areas and in all types of farming. A stigma is still attached to the idea in some parts of the United States, just as in other parts a farmer is considered shortsighted if he passes up an opportunity to become a tenant and thus increase the size of his operation. As a general rule, tenancy is least successful on farms with major long-term capital investments in such things as orchards, vineyards, pasture improvements, or herds of livestock. The owner does not want to run the risk of having his investment destroyed by a careless tenant, whereas a tenant would be understandably reluctant to invest in long-term improvements, which will not yield a return very quickly, unless he is secure in his tenure, or unless he has legal assurance that he will be compensated for the unused improvements if the landlord breaks his lease.

The worst types of farm tenancy reduce the tenant to the status of a mere laborer. With the exception of slavery and sharecropping on cotton and tobacco farms in the South, and the use of migrant farm workers on intensive fruit, truck, and sugar beet farms, most parts of the United States have been fortunate enough (thanks in large measure to the family farm ideal) to have escaped the corrosive effects of large landed estates, subservient tenants, and a faceless laboring class of peons,

---

[18] Howard F. Gregor, "Industrialized Drylot Dairying: An Overview," *Economic Geography,* Vol. 39 (1963), 316.

serfs, and slaves.[19] In many parts of Europe and Latin America these have sparked demands for agrarian reform, or even revolution.

The Cotton Plantation in the American South

T. Lynn Smith has painted a vivid picture of the debasing effects of large-scale estate, or plantation, agriculture.[20] Society becomes stratified, with an elite who believe that manual labor and hard work are degrading, and a rural proletariat whose repetitive chores stunt the development of their intellects and personalities. Stratification restricts social mobility, and tends to breed distinction, conflict, and struggle between the classes, but all classes are conservative in their resistance to innovation, progress, and change. The workers have no reason to believe that they can raise their low level of living by hard work or thrift, and the elite are reluctant to save for investment in improvements which are likely to be misused or neglected unless they are very carefully supervised. These problems are magnified when the barriers of class are reinforced by differences in race, language, or religion.

Many of the unattractive features which Smith has described so vividly are exemplified in the cotton plantations of the southeastern United States.[21] These large landholdings, which have proven remarkably persistent under changing social and economic conditions, also show how a sytem of land tenure can influence the look of the land. The term "plantation," as it was first used in the Chesapeake Bay tobacco country, applied simply to a clearing, or "planting," within a tract of wooded country, and not to the landholding itself.[22] In fact, the British Colonial Office was originally known as the Plantation Office, and the term "plantation" was not used to describe a tropical estate until the eighteenth century, but it quickly came to be synonymous with a method of production from the land and a way of life.

Plantation agriculture may be defined as the use of cheap (often slave) unskilled labor and primitive hand methods on large landholdings to produce a tropical cash crop for export to the middle latitudes. In the New World it apparently was tried first in the sugar-producing districts of northeastern Brazil in the sixteenth century, but it soon spread to the sugar islands of the West Indies, and thence to the tobacco districts of Virginia and Maryland. Indentured servants, rather than slaves, were

---

[19] Duilio Peruzzi, "The Mezzadria: A Decaying System of Land Tenure and Management in Italy," in E. S. Simpson, ed., *Agricultural Geography I. G. U. Symposium*, Research Paper No. 3 (Liverpool: University of Liverpool Department of Geography, 1965), pp. 49–54.

[20] T. Lynn Smith, *The Sociology of Rural Life*, 3rd ed. (New York: Harper & Row, 1953), pp. 313–19.

[21] T. J. Woofter, Jr., *Landlord and Tenant on the Cotton Plantation*, Research Monograph 5 (Washington, D.C.: Works Progress Administration, 1936); Harald A. Pedersen and Arthur F. Raper, *The Cotton Plantation in Transition*, Bulletin 508 (State College, Miss.: Mississippi Agricultural Experiment Station, 1954); and Merle Prunty, Jr., "The Renaissance of the Southern Plantation," *Geographical Review*, Vol. 45 (1955), 459–91.

[22] Ralph H. Brown, *Historical Geography of the United States* (New York: Harcourt Brace Jovanovich, 1948), p. 59.

used in the early years on the tobacco plantations, but as soon as they had worked out their period of indenture they obtained land of their own and began to compete with their former masters. The planters soon shifted to the use of Negro slaves imported from Africa, and slavery became an integral part of the plantation system which spread to the rice and indigo districts of South Carolina and the sugar cane areas of Louisiana.

The plantation system flourished with cotton, which became the staple crop of the South at the end of the eighteenth century. Although large plantations were never the mainstay of agriculture in the region, their customs and ideology set the patterns for smaller units, because the wealth and influence of the large plantation landowners enabled them to assume political and economic leadership in a society which remained essentially rural. Even today the ownership of land confers more social status in the South than it does in other parts of the nation; the word "plantation," like "ranch," is value-loaded, and carries great prestige within the region, if not nationally. The strength of this prestige is illustrated by the fact that Prunty has coined the term "neoplantation" to describe a large, mechanized farm with a hired labor force in those parts of the South which have a plantation tradition, and he has classified the quail-shooting preserves and winter resorts of wealthy northern industrialists as "woodland plantations" if they are managed for production of forest products.[23]

The focus of the antebellum plantation was the "big house" of the planter and his family, which was seldom quite as fancy as Hollywood would have us believe. Not far away were "the quarters," rows of cabins for the slaves, plus implement and tool sheds, the mule barn, the blacksmith shop, and perhaps a cotton gin. Most plantations were on level to undulating land. They were laid out in large fields, which were worked by gangs of Negro slaves and mules under the watchful eye of the planter or his overseer. Cotton was grown on most of the cropland, and many plantations produced little or none of their own food or feed. Before the Civil War, in fact, the city of Cincinnati was nicknamed "Porkopolis," because its prosperity depended on processing the hogs produced on the farms of the emerging Corn Belt to the north, and shipping the pork to southern plantations.

The emancipation of slaves worked great changes in the plantation system.[24] After the Civil War the planter still owned the land, but he had no one to work it, and the freed slaves had no land. From this dilemma emerged the infamous system known as sharecropping; the planter provided the land, the ex-slave provided the labor, they split the cost of seed and fertilizer, and they shared the crop which resulted. The ex-slave wanted a piece of land of his own, so the planter divided the large fields

    [23] Prunty, "The Renaissance of the Southern Plantation," footnote 21; and Merle Prunty, Jr., "The Woodland Plantation as a Contemporary Occupance Type in the South," *Geographical Review*, Vol. 53 (1963), 1–21.
    [24] Prunty, "The Renaissance of the Southern Plantation," footnote 21, has superb maps illustrating how the layout of plantations has changed.

of the old plantation into smaller sharecropper units of 20 to 40 acres, which was about as much land as one family could handle (Fig. 5–4). The ex-slave wanted the privacy of his own house on his own land, so the planter razed the old slave quarters and built new sharecropper cabins, which were scattered through the cotton fields. The ex-slave wanted to be able to use his mules for travel as well as for work, so the planter let him have them.

Although the planter had to make many concessions, the one thing he did not relinquish was control of the work that was done on his land. This control was just as tight as he could make it and still keep his crop-pers from "skipping out" on him before they had harvested the crop. Many plantations had a bell which was rung when it was time to get up in the morning, when it was time to begin work, and when it was per-missible to stop. The planter decided when it was time to plant cotton in the spring, when it was time to get out the hand hoes and start chop-ping out the grass and weeds which were growing up between the young plants, when it was time to start picking the ripe crop, and when and where it would be sold. The planter or his agent were constantly making their rounds to inspect the work and supervise details, and the share-cropper was essentially a laborer who made no management decisions.

Sharecropping involved a great deal of indebtedness. In the spring the cropper expected the planter to advance him enough money to pay for his share of the cost of seed and fertilizer, and when the crop was in the ground he expected the planter to lend him "furnish" to buy food and other necessities. "Furnish" was always a headache, and some plantations had commissary stores at which croppers were expected to obtain their needs. At settlement time in the fall the planter subtracted all these ad-vances from the cropper's share before paying him off, but often the planter himself was short of cash, and he had to borrow money to cover his costs during the growing season. In a bad year the planter lost not only his share of the crop, but also the money he had advanced to his croppers.

A good planter took a keen paternal interest in his croppers, and he was reluctant to replace them by machines until he was forced to do so. Many planters, in fact, did not mechanize their operations until World War II, when the massive migration of former sharecroppers to city jobs created serious labor shortages. Mechanization has made major changes in the plantation landscape. The old sharecropper "cabin in the cotton patch," often an unpainted, unscreened, frame shack with primitive water supply and sanitation facilities, is rapidly disappearing (Fig. 5–5). Some have been razed, and many of those which remain are used only for storage. New and better houses have been built along the highway for the tractor drivers and other wage earners who form the much reduced labor force, although sometimes there are mutterings about "quarters" and slavery times if the houses are placed too close together.

Small fields have been combined into larger ones so that the new machines can be used at maximum efficiency, and shiny new tractor sta-tions and implement sheds have been built to shelter them. Government

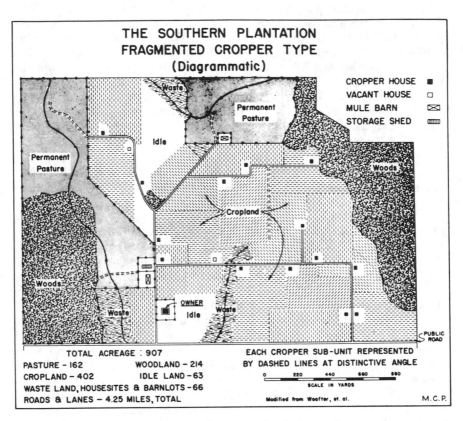

FIG. 5–4. *Diagram of a cotton plantation divided into sharecropper units of 20 to 40 acres each. Reproduced by permission of the author and publisher from Merle Prunty, Jr., "The Renaissance of the Southern Plantation,"* Geographical Review, *Vol. 45 (1955), 467. Copyright by the American Geographical Society of New York.*

acreage control programs have restricted cotton acreages and encouraged many planters to experiment with new crops, such as soybeans, or to convert some of their land into improved pastures for livestock. The modern planter is very fortunate, in fact, because he already owns a large acreage of land at a time when farmers all over the nation are trying to, and need to, enlarge their holdings to a size comparable to his. Apart from its history, and the fact that the labor force may be predominantly Negro, the modern plantation is not very different from any other large farm in the United States which depends upon a hired labor force.[25]

[25] Howard F. Gregor, "The Plantation in California," *Professional Geographer,* Vol. 14, No. 2 (March 1962), 1–4.

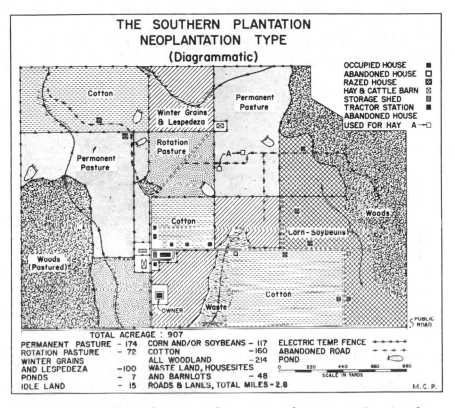

THE SOUTHERN PLANTATION
NEOPLANTATION TYPE
(Diagrammatic)

OCCUPIED HOUSE
ABANDONED HOUSE
RAZED HOUSE
HAY & CATTLE BARN
STORAGE SHED
TRACTOR STATION
ABANDONED HOUSE
USED FOR HAY   A→□

Cotton

Winter Grains
& Lespedeza

Permanent
Pasture

Rotation
Pasture

Permanent
Pasture

A

Cotton

Woods

Corn-Soybeans

Woods
(Pastured)

Cotton

OWNER   Waste

PUBLIC
ROAD

TOTAL ACREAGE : 907

PERMANENT PASTURE – 174	CORN AND/OR SOYBEANS – 117	ELECTRIC TEMP. FENCE
ROTATION PASTURE – 72	COTTON – 160	ABANDONED ROAD
WINTER GRAINS	ALL WOODLAND – 214	POND
AND LESPEDEZA – 100	WASTE LAND, HOUSESITES	
PONDS – 7	AND BARNLOTS – 48	
IDLE LAND – 15	ROADS & LANES, TOTAL MILES – 2.8	

0   220   440   660   880
SCALE IN YARDS

M.C.P.

FIG. 5–5. *Diagram of the former sharecropper plantation in Fig. 5–4 show-ing how it was reorganized after World War II. The sharecropper "cabins in the cotton patch" have been razed, new and better houses have been built for the wage workers who remain, fields have been enlarged and fenced, and new crops and livestock have been introduced. Reproduced by permission of the author and publisher from Merle Prunty, Jr., "The Renaissance of the Southern Plantation," Geographical Review, Vol. 45 (1955), 482. Copyright by the American Geographical Society of New York.*

CHAPTER 6 *farm enlargement*
and farm management

In the United States the family farm of yesterday is too small for today, and today's family farm will probably be too small tomorrow. Since the Second World War successful farmers in all parts of the nation have been aware of the necessity of enlarging the size of the farm business. Planters in the South were fortunate, because they have been able to assemble adequately large acreages merely by reorganizing the land which they already own, but elsewhere the farmer who has wished to enlarge his operation has had to buy or rent the additional land he needed.

Most of the pressure for farm enlargement has resulted from the replacement of the horse by the internal combustion engine, a process which did not really begin to get under way until the late 1930s.[1] In 1926, for example, Searle could write that "it is not likely that the horse will ever be entirely displaced; at least one team will be necessary on most farms."[2] Even as late as 1948 De Graff and Haystead seem to have believed that there was still serious debate about the relative merits of the horse and the tractor, because they devoted no less than 11 pages to a discussion of the subject.[3] Today, however, in addition to a tractor, and perhaps a self-propelled combine harvester and/or corn picker as well, an efficient farmer must have a second tractor for hauling grain during harvest, extra power during critical work periods, and miscel-

[1] Austin Fox, *Demand for Farm Tractors in the United States: A Regression Analysis,* Agricultural Economic Report No. 103 (Washington, D.C.: U.S. Department of Agriculture, 1966), p. 29.
[2] C. F. Searle, "Horse Production Falling Fast in U.S." *Yearbook of the U.S. Department of Agriculture, 1926* (Washington, D.C.: U.S. Department of Agriculture, 1927), p. 438.
[3] Herrell De Graff and Ladd Haystead, *The Business of Farming* (Norman, Okla.: University of Oklahoma Press, 1948), pp. 72–83.

laneous jobs around the farm.[4] And this new and larger machinery is enabling him to work more land in less time.

In the old days one man with horse-drawn equipment could not handle more than about 80 acres of Corn Belt land by himself, and he needed help from his son or a hired man if he wished to farm a larger acreage; the maximum size of a father-and-son, family farm operation was fairly close to the 160 acres allotted under the Homestead Act.[5] In contrast, Van Arsdall and Elder estimated that one man with 1969 machinery should be able to handle 580 acres of land in the cash-grain country of central Illinois if he used four-row equipment, and 850 acres if he had eight-row equipment; they concluded that the most profitable farm in this area would be a two-man family farm, with a total capital investment of $900,000, using eight-row equipment on 1,640 acres of land.[6] In the corn-/hog farming country of western Illinois they believed that the optimal farm would be a two-man operation of only 720 acres with a total capital investment of only $330,000.

Part Ownership as a Strategy for Farm Enlargement

How does a farm quadruple in size within a few decades? The answer seems to be that the smaller farmers are dropping out of the race, and the larger farmers are taking over their land, either by purchasing it or, in many cases, by renting it. Thirty-one of a sample of 227 southeastern Pennsylvania farms went out of business between 1960 and 1965; 19 of these 31 had been incorporated in other farms, 11 by lease and eight by sale.[7] A study in Indiana concluded that most farmers increased their acreage by renting additional land.[8] The part owner, who owns part of his land and rents the rest, now stands on the top rung of the tenure ladder, in many cases full ownership represents a retreat to a smaller volume of business and a smaller income with the encroachment of age.

The man who already owns his land, and has surplus labor, machinery, and buildings, can afford to rent a smaller parcel of land, and can pay a higher rent, than a tenant who must depend on the rented land for his entire income. The man who has land to lease prefers a tenant who already owns some land, in the assumption that he will be

[4] Paul E. Strickler, *Machines and Equipment on Farms, With Related Data, 1964 and 1959*, Statistical Bulletin No. 401 (Washington, D.C.: U.S. Department of Agriculture, 1967).

[5] Wayne E. Kiefer, *Rush County, Indiana: A Study in Rural Settlement Geography*, Geographic Monograph Series, Vol. 2 (Bloomington, Ind.: Indiana University Department of Geography, 1969), p. 134.

[6] Roy N. Van Arsdall and William A. Elder, *Economies of Size of Illinois Cash Grain and Hog Farms*, Bulletin 733 (Urbana, Ill.: Illinois Agricultural Experiment Station, 1969), p. 25.

[7] Glenn A. Zepp and Robert H. McAlexander, *Farm Adjustments in Southeastern Pennsylvania, 1960–1965*, Progress Report 274 (University Park, Pa.: Pennsylvania Agricultural Experiment Station, 1967), pp. 1–3.

[8] V. Harrison, C. Langbehn, and L. M. Eisgruber, "Capital Accumulation and Resource Adjustments," *Economic and Marketing Information for Indiana Farmers*, June 1969, pp. 1–4.

more responsible, and that he will not require a house to live in nor any farm buildings other than a few simple shelters; a farmstead, whether or not it is occupied, has such an attraction for the tax assessor that the owner prefers to raze any buildings which are not necessary.

Part ownership may also be a useful strategy for keeping the better land in production in areas of very mixed soils and surface features. A rectangular landholding in such an area may contain some very good land, as well as some land which is not worth farming. The owner may use his good land as a base of operations, and "put together" a farm of adequate size by renting land from other landowners in a similar situation, or he may take an off-farm job and rent his better land to another farmer. This strategy has been especially appealing in parts of Appalachia, where land of good quality is commonly found in close proximity to very poor land, and where reluctance to sell the old family place is so strong that it is quite difficult to purchase enough land for a farm large enough for modern mechanized operations.

An American farmer who wishes to expand the size of his operation often has to travel a considerable distance to find a suitable tract of land which he can rent or buy. Farm expansion, even at a modest level, commonly requires the operation of noncontiguous tracts of land, and in many cases the process of farm enlargement actually appears to be a process of farm fragmentation. Two of every five farms in nine sample Minnesota townships consisted of two or more noncontiguous parcels of land, often separated by several miles.[9] More than half of the farms of one township in Lincoln County consisted of noncontiguous units (Fig. 6–1). A quarter of the land was comprised of noncontiguous tracts, which ranged in size from 40 to 320 acres, and were separated from the farmstead by road distances of one to 22 miles.

Farm operating units consisting of noncontiguous tracts are no new phenomenon, nor are they unique to Minnesota. Many early farmers on the open prairie of Story County, Iowa, owned or rented stream-side woodlots, sometimes as much as 12 miles from the farmstead.[10] A quarter of the farmland in a Nebraska community in the 1930s consisted of noncontiguous tracts, and dispersed tracts of land under a single ownership were common in the Oxnard area of California.[11] Farms consisting of noncontiguous tracts have been common both on claypan areas in the southern part of Illinois and in the cash-grain farming areas of the Grand Prairie in the east center.[12]

[9] Everett G. Smith, Jr., "Road Functions in a Changing Rural Environment," unpublished doctoral dissertation, University of Minnesota, 1962, pp. 71–75.

[10] Leslie Hewes, "Some Features of Early Woodland and Prairie Settlement in a Central Iowa County," *Annals*, Association of American Geographers, Vol. 40 (1950), 53.

[11] Robert Diller, *Farm Ownership, Tenancy, and Land Use in a Nebraska Community* (Chicago: University of Chicago Press, 1941); and Howard F. Gregor, "A Sample Study of the California Ranch," *Annals*, Association of American Geographers, Vol. 41 (1951), 294.

[12] A. J. Cross and J. E. Wills, *Organization and Operation of Farms in the Claypan Area of Southern Illinois*, Bulletin No. 579 (Urbana, Ill.: Illinois Agricultural Experiment Station, 1954); and Van Arsdall and Elder, *Economies of Size*, footnote 6, p. 48.

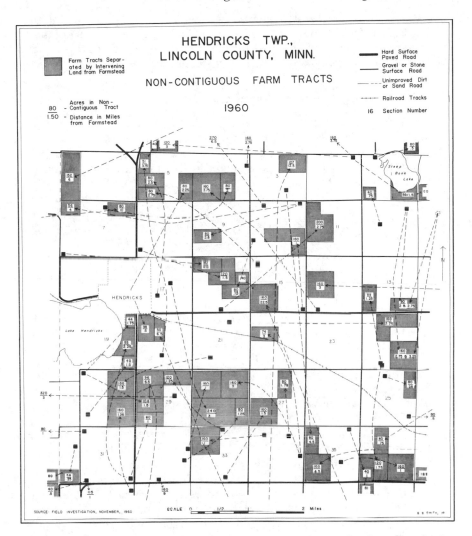

FIG. 6–1. *Farms consisting of noncontiguous tracts of land in Hendricks Township, Lincoln County, Minnesota. Reproduced by permission of the author from Everett G. Smith, Jr., "Road Functions in a Changing Rural Environment," unpublished doctoral dissertation, University of Minnesota, p. 64.*

The ability of an individual to farm physically separated tracts of land is closely related to the nature of the local rural road system, which farmers must use in moving their equipment from one tract to another. In Minnesota many local roads, which at first glance might be considered nonessential because no dwellings front on them and they carry little traffic, are actually vital because they provide the shortest routes between farmsteads and noncontiguous tracts of land.[13] The section-line

[13] Smith, "Road Functions in a Changing Rural Environment," footnote 9, pp. 51 and 77.

roads of the Middle West, which crisscross the countryside at one-mile intervals, have probably facilitated part owner farming of noncontiguous tracts because they have enabled farmers to travel and operate farmland in almost any direction from their farmsteads with equal ease.[14]

Off-Farm Residence of Farmers

The increasing number of farms which consist of separated tracts of land raises the question of whether the isolated farmstead itself may be an anachronism, because it is commonly believed that fragmentation of properties tends to encourage farmers to live in central nucleated settlements.[15] During the early 1960s, however, a series of investigations of small central places scattered through the Middle West found scant evidence in support of this belief, and a study of 340 hamlets in Nebraska discovered that only one of every eight household breadwinners was a farmer.[16]

Perhaps the most definitive answer to the question of whether dispersed farm tracts favor central residence is provided by the work of Kollmorgen and Jenks on suitcase and sidewalk farming on the Great Plains.[17] A sidewalk farmer, according to their definition, lives in town, and a suitcase farmer lives 30 miles or more from the border of the county in which his farmland is located. Sidewalk farmers seem to feel compelled to justify their decision to live in town, and little justification for such a decision can be found in a densely settled rural area which has more or less adequate farm housing, roads, schools, utility services, and opportunities for social contacts and nonfarm employment. In an eastern North Dakota county, for example, only 14 percent of the farmers in 1953 were sidewalk farmers.

By way of contrast, in north central Montana 30 percent of the farmers in Toole County in 1955 were sidewalk farmers. The early settlement of Toole County was followed by a series of dry years, and much of the land was abandoned or became tax delinquent. Subsequently the risk of farming has been reduced considerably by new technology and new farming methods, which have been introduced, for the most part, since 1940, and the farmers who remained in the county have expanded their operations by buying tax delinquent land or by renting land from absentee owners who fled in the early years. Much of this land lacks

---

[14] Ibid., footnote 9, p. 114.

[15] T. Lynn Smith, *The Sociology of Rural Life*, 3rd ed. (New York: Harper & Row, 1953), p. 216.

[16] John Fraser Hart, "Vermontville and Augusta: A Study of Two Michigan Villages," *Papers of the Michigan Academy of Science, Arts, and Letters*, Vol. 49 (1964), 420; John Fraser Hart, Neil E. Salisbury, and Everett G. Smith, Jr., "The Dying Village and Some Notions about Urban Growth," *Economic Geography*, Vol. 44 (1968), 343–49; and Albert J. Larson and A. P. Garbin, "Hamlets: A Typological Consideration," *Sociological Quarterly*, Vol. 8 (1967), 531–37.

[17] Walter M. Kollmorgen and George F. Jenks, "Suitcase Farming in Sully County, South Dakota," *Annals*, Association of American Geographers, Vol. 48 (1958), 27–40; and Walter M. Kollmorgen and George F. Jenks, "Sidewalk Farming in Toole County, Montana, and Traill County, North Dakota," *Annals*, Association of American Geographers, Vol. 48 (1958), 209–31.

adequate houses and roads, and a large part of the county does not even have good drinking water. The man who is farming several tracts dispersed over a considerable distance would not reduce his travel very much if he lived on any one of them, and by living in town he and his family have access to the amenities of schools, stores, health care, a supply of drinking water, and the possibility of nonfarm employment.

An even more extreme example of the possible impact of farmland dispersal on the place of residence of the farmer is provided by Sully County, in central South Dakota, where less than 10 percent of the farmers in 1952 were sidewalk farmers, but about 15 percent were suitcase farmers, most of whom lived several hundred miles to the south in the Winter Wheat Belt. The county has a history of severe and prolonged droughts, which has instilled an attitude of caution in the local farmers, and the traditional crop and livestock farming system in the county is diversified as a hedge against drought hazards and price fluctuations.

The suitcase farmers grow only a single crop, wheat, but they have introduced a new kind of diversification, geographical diversification, by growing that one crop in several different areas. They know that drought rarely hits all wheat-producing districts in the same year, and if they have planted wheat in several different districts they are likely to get a crop in at least one.[18] Furthermore, many of these men are combine operators who have been accustomed to following the wheat harvest northward, and they have realized that their expensive plowing and planting equipment could be kept busy for a longer time each year if it also followed the northward advance of wheat farming operations. Sully County, on the main north-south highway long used by combine operators, was a logical place to acquire land when they began to diversify geographically, although they visit the county only briefly each year when farm operations are necessary.

The Management of Larger Farms

Suitcase farming, as it is practiced in Sully County, provides an illustration of the way in which new concepts of farm management are being developed to replace traditional methods. Changes in management systems and management thinking are inevitable as farms grow larger, although the precise form that these changes will take is still far from clear. The inevitability of changes, and attendant uncertainty as to what those changes will be, is one of the reasons why it is so exciting to be a student of the American agricultural scene in the last quarter of the twentieth century. Although I do not anticipate the development of "factories in the fields," whatever that phrase may mean, I do believe that the next few decades might well witness changes in farming whose

[18] Clarence W. Jensen and Darrel A. Nash, *Farm Unit Dispersion: A Managerial Technique to Reduce the Variability of Crop Yields*, Bulletin 575 (Bozeman, Mont.: Montana Agricultural Experiment Station, 1963).

scale could rival those which took place in manufacturing in the latter part of the eighteenth century.

A generation ago farming was an art and a way of life; today, in response to technological innovations and the rapid increase in scale of operations, the successful farm has to be a modern business operation which uses the latest and most up-to-date techniques, both in production and in management. As farms become larger, they will have to become more specialized, and the farmer, like any other businessman, will have to concentrate on doing what he can do best, most efficiently, and most competitively. The family farm, which used to produce most of what it needed and need most of what it produced, will have to specialize on the product to which it is best adapted, and buy everything else.

The replacement of horses by tractors has relieved the farmer of the necessity of growing feed and pasture, but it has also forced him to increase his volume of sales so that he can pay for gas, oil, and the tractor itself. Better roads and easier access to stores have reduced the need for the farm family to produce its own food, and on many farms the garden and orchard have long since disappeared; most farmers today buy their food in the same supermarkets where you and I buy ours. Trucks and better roads have also facilitated the transportation of feed from one farm to another, and thus have permitted the separation of feed production and livestock feeding; one farmer can specialize in producing feed, and another in feeding it to livestock.

CORPORATE FARMING. Despite some of the dire predictions which have been made, I do not believe that any of these developments point inevitably in the direction of corporate farming, and I suspect that the threat of the corporate farm has been grossly exaggerated. The corporate farm alarmists appear to have confused "big-ness," which does seem inevitable in agriculture, with "corporate-ness," which almost certainly is not, because corporate organization is not a prerequisite to capitalizing on any of the efficiencies of size in coordinating and integrating production, processing, and distribution. In fact, the bills forbidding farms to grow larger or to incorporate which have been introduced in some state legislatures are faintly reminiscent of Canute's command to stop the tide from rising, or the bill to change the value of *pi* to 3.00, because the good legislator could never remember 3.1416.

A recent survey by the Department of Agriculture revealed that only about 11,000 farms in the United States are owned by corporations, and four-fifths of these corporations are individual or family affairs. As farms grow larger and the amount of capital required for modern farming continues to increase, one would expect that more and more farmers would adopt the corporate form of business organization to improve their borrowing position, to facilitate the accumulation of capital, to limit liability in case of failure, to gain tax advantages, and to ease the transfer of farm assets from one generation to the next.

PROFESSIONAL FARM MANAGERS. Although we have little evidence of any such trend as yet, I think it is safe to assume that the acreage of land and the number of farms which are operated by professional farm managers will increase. A small fraction of our farmland is already in the hands of banks, insurance companies, elderly persons, or city heirs who require the services of a professional farm manager because they have neither the inclination nor the competence to manage it themselves. The professional manager, or farm management service, has the job of finding tenants, planning a cropping and stocking program, preparing periodic reports and accounting statements, arranging the marketing of products, and doing everything else that an ordinary owner would do for himself. I suspect that professional management can be successful only if the owner is interested solely in the income he receives from the farm, and has little more physical contact with or interest in his land than a shareholder might have with a factory in which he holds stock.

VERTICAL INTEGRATION. I might use vertical integration in poultry farming as an example of a new farm management system which seems to be working fairly well at present, but one which appears doomed to failure in the long run, because most farmers do not like it. A poultry farmer can fatten day-old baby chicks to broiler size in about nine and a half weeks, but he is only one link in a chain connecting the hatchery which produces the chicks, the milling company which buys the grain and prepares the feed, the processing plant where broilers are dressed for market, and the chain of retail outlets where they are sold to consumers.

Formerly many poultry farmers bought their own chicks and feed, often with borrowed money, but the profit margin on broilers is so low, and the market fluctuates so rapidly, that some of them went broke, and most have become unwilling to risk their own capital in the poultry business. The vertical integrator (often a feed dealer or processing company) stepped into the picture by forging a complete chain of chick hatcheries, grain elevators, feed mills, processing plants, and retail outlets. He can afford to provide the poultry farmer with the necessary risk capital, but in return he retains the right to make all management decisions. The poultry farmer has a contract which guarantees him a fixed price for the birds he feeds, and he likes this sure labor income, but he intensely resents the fact that an outsider can issue him orders on his own farm, and many poultry farmers have told me that they would like to get out of the broiler business as soon as they can afford to do so.

THE CITRUS BUSINESS. A more successful new management system has evolved in the citrus business, in large part because ownership of a citrus grove is a form of investment. Most grove owners are local professional people who have made enough money to invest in a grove, or well-to-do residents of distant cities who are attracted by the idea of investing in a citrus grove in sunny Florida, and neither group has the time or inclination to do their own work. Furthermore, most groves are too small to justify the expense of necessary equipment, and their management is entrusted to professional caretakers, or to the production departments of citrus-packing companies, which have been drawn into the business largely by default after they concluded that they could obtain fruit of the quantity and quality they needed only if they produced it themselves.

A typical production department of a citrus-packing company would have charge of several hundred groves covering thousands of acres. Its headquarters has enormous shelters for trucks, tractors, and other mobile equipment, and smaller sheds for storing insecticides, fertilizer, and gasoline and oil. A radio tower looms high above it, because all mobile equipment has two-way radios with a range of 50 miles or more, and all operations are directed by radio. The production department is prepared to fertilize, cultivate, spray, prune, plant, and perform all other necessary grove operations. The grove owner is billed for each operation, but these charges are levied against the credit he receives for his fruit after it has been picked, and his only real contact with his grove is the annual check. "I've worked for some of my grove owners for over 20 years now," said one production manager, "but I have never once knowingly set eyes on them. Of course any car with a New York license which is parked beside the grove could have the owner inside, but I have no way of telling."

The Dry Lot

DRY-LOT FEEDING.   One of the most interesting recent developments in agricultural management has been the rapid growth of specialized livestock feeding operations in the West, and the expansion of this system into the East. The use of grain, mainly corn, to fatten livestock has been important in the Corn Belt ever since Federal regulations reduced the profit to be made from distilling corn into spirits and encouraged farmers to market their grain on the hoof.[19]

The fattening or finishing of livestock, known as feeding, became increasingly important as settlement spread westward into the drier grasslands, which could produce bone and hide, but not the kind of meat demanded by the American consumer. For half a century or more, lean animals from the range country, known as "feeder cattle," have been shipped to farms in the western Corn Belt, where they have been fattened for slaughter. The Corn Belt farmer buys his cattle in October or November, after the harvest is in, feeds them out to slaughter weight during the winter, and sends them off to the stockyards in spring or early summer before his crop work begins again.[20] Ideally the cattle are placed on feed at a weight of around 650 pounds, are fed 120 to

---

[19] Justinian Caire, *Cattle Feeding and Its Place in Twelfth District Agriculture,* Supplement to the Monthly Review (San Francisco: Federal Reserve Bank, January 1953), p. 7.
[20] Robert M. Finley and Ralph D. Johnson, *Changes in the Cattle Feeding Industry in Nebraska,* Bulletin 476 (Lincoln, Neb.: Nebraska Agricultural Experiment Station, 1963), p. 4.

150 days, and are sold at a weight of around 1,050 pounds, for an average daily gain of 2 to 2.5 pounds, or slightly more.[21]

Cattle feeding has spread to new areas since World War II, and it has become an increasingly mechanized, specialized, assembly line method of production which is concentrated in a smaller number of larger operations. Feeding 100 steers was just about a full-time job for one man in the 1920s, but today, with labor-saving machinery and pushbutton control, one man should be able to handle 500 to 1,000 or more.[22] The growth of population, a rising demand for meat, and the availability of feed from irrigated farming areas have produced a rapid growth of cattle feeding in the West, but land in irrigated areas is too precious to be used for pasture, and many of the cattle are fattened on dry lots.

A dry lot is a piece of bare ground, without grass but provided with water and troughs where cattle can be fed. The individual pens, or "corrals," are enclosed by stout fences. The feed troughs run along one side of the corrals, and outside the fence so that they may be filled mechanically from a self-unloading truck driving slowly down the lane between them. A single corral holds 40 to 80 animals and covers half an acre, but for maximum profit a feedlot should have a capacity of at least 1,500 head; in 1964 the West had more than 1,000 feedlots which equalled or exceeded this capacity, and the largest one had a capacity of 34,000 head.[23]

Corn, grain sorghum, and barley are the principal grains fed, and alfalfa hay, corn silage, and hay provide most of the necessary roughage, but the cost of feed is such a major factor in the success of the feedlot that the operator will feed almost anything he can purchase cheaply and the cattle are able to get down. Cottonseed hulls and cake are staple feeds in southern irrigated areas, as are sugar beet tops and pulp in northern areas, but the list of feeds also includes spent mash from distilleries, citrus pulp from juice concentrating plants, grape pumice from wineries, almond hulls, peanut meal, olive meal, pear pulp, offal from vegetable and fruit canning plants, and surpus raisins, prunes, potatoes, carrots, and cantaloupes.[24]

Some of the most spectacular developments in dry-lot feeding have occurred in California, which in 1960 had eight of the 16 leading counties in the United States in total number of cattle and calves on farms, and six of the ten leading counties in number of cattle and calves sold alive.[25]

[21] Ronnie L. Burke, *Characteristics of Beef Cattle Feedlots: California, Colorado, Western Corn Belt,* Marketing Research Report No. 840 (Washington, D.C.: U.S. Department of Agriculture, 1969), pp. 21, 24, and 29.

[22] R. T. Burdick, *Cost of Cattle and Lamb Feeding in the Northern Colorado Irrigated Area,* Bulletin 439-A (Fort Collins, Col.: Colorado Agricultural Experiment Station, 1955), p. 12.

[23] *Number of Feedlots by Size Groups and Number of Cattle Marketed, 1962–1964,* Statistical Reporting Service 9 (Washington, D.C.: U.S. Department of Agriculture, 1966).

[24] Burke, *Characteristics of Beef Cattle Feedlots,* footnote 21, p. 37.

[25] John Fraser Hart, "Meat, Milk, and Manure Management in the West," *Geographical Review,* Vol. 56 (1966), 118–19.

The state's leading cattle-feeding area was the Imperial Valley, in the southern part, where the number of cattle tripled between 1950 and 1960. Most of the beef animals in the Imperial Valley are fattened on some 20 feedlots, each of which can handle more than 3,500 head at a time. A few large yards have an annual turnover of about 2.4 times their capacity.[26]

In another important feeding area, the High Plains of western Texas, cattle feeding began to blossom only in the 1960s, in response to a reduction in the cost of shipping cold meat to market, plus an increase in the cost of shipping feed grains and cattle.[27] Greater efficiency in shipping cold meats enabled the meat packing industry to decentralize, and new packing plants were built near the source of cattle; during the 1960s the High Plains added eight new plants, with a total capacity of 2.2 million head, to the 1960 inventory of 12 plants with a total capacity of only 0.4 million. The number of cattle on feed in the area shot from 100,000 in 1960 to 950,000 in 1969 simply by retaining Texas feeder cattle which formerly had been shipped to feedlots in other areas. Abundant feed was available for them in the form of milo maize, a grain sorghum which local farmers had begun to grow on a large scale in the late 1950s as the government reduced cotton acreage allotments.

Two studies in California indicate that the availability of feed does not necessarily provide an adequate base for a profitable cattle-feeding operation. In the Sacramento Valley, where rice acreages had been restricted by government controls, potential management income on a thousand-acre farm could have been increased from $10,000 to $40,000 or more a year in the early 1960s by adding a cattle-feeding program.[28] In the Imperial Valley, however, a well-managed cash crop farm of 480 acres should already have had management income of around $30,000 in 1960, and the addition of a feeder cattle operation would have required an initial capital investment of at least $100,000, the development of new management skills and business contacts, and a willingness to accept considerably greater financial risks.[29]

DRY-LOT DAIRY FARMING.   The dry-lot system of intensive livestock management need not be restricted to feeding beef cattle. The Los Angeles area, for example, has become an intensive dry-lot dairy farm-

[26] Robert A. Kennelly, "Cattle-Feeding in the Imperial Valley," *Yearbook of the Association of Pacific Coast Geographers*, Vol. 22 (1960), 50–56.
[27] Charles M. Wilson, "The Cattle Feeding Industry in the High Plains," *Business Review*, Federal Reserve Bank of Dallas, July 1969, pp. 3–9; Gene L. Swackhamer and Blaine W. Bickel, "Cattle Feeding in the Tenth District: Development and Expansion," *Monthly Review*, Federal Reserve Bank of Kansas City, April 1970, pp. 13–22; and R. A. Dietrich, J. R. Martin, and P. W. Ljungdahl, *The Capital Structure and Financial Management Practices of the Texas Cattle Feeding Industry*, Bulletin 1128 (College Station, Tex.: Texas Agricultural Experiment Station, 1972).
[28] James A. Pettit, Jr., and Gerald W. Dean, *Economics of Farm Feedlots in the Rice Area of the Sacramento Valley*, Bulletin 800 (Davis, Calif.: California Agricultural Experiment Station, 1964), p. 1.
[29] H. O. Carter, G. W. Dean, and P. H. Maxwell, *Economics of Cattle Feeding in Imperial Valley Field Crop Farms*, Bulletin 813 (Davis, Calif.: Califonia Agricultural Experiment Station, 1965), pp. 48–49.

ing area, with large herds on remarkably restricted acreages. An area of 60 square miles near Chino, California, had approximately 400 dairy farms with 160,000 cows in 1971; in contrast, a normal farm in the better dairying areas of Wisconsin would have only 75 to 100 cows. The most densely populated areas in California had more than 50 cows to the acre. Some dry-lot dairy farms have small grass plots to provide drainage areas for the large amounts of water they use, but many have no land other than the bare corrals. The principal building on a dry-lot dairy farm is a "walk through" milking barn, with pipelines which carry the milk to cooling tanks in the adjoining milk house.[30] The dry-lot dairy farmer, of necessity, emphasizes quality cattle of high productivity and is quick to cull low producers. His location near the port of Los Angeles enables him to use concentrated feeds imported from both domestic and foreign sources, including copra meal from the Philippines and pineapple residues from Hawaii.

In an attempt to escape the problem of inflated land values and escalating taxes on a rapidly growing fringe, between 1955 and 1957 three separate areas southeast of Los Angeles were incorporated solely for the purpose of preserving them as dairy farming areas.[31] In 1960 Dairy Valley had 53,000 cows on 241 farms, Cypress had 14,000 cows on 50 farms, and Dairyland had 8,300 cows on 32 farms. These places still have two major problems: conflicts with nearby residents over odors, flies, and animal and dairy noises; and the building up of adjacent open areas which the dairymen had previously been using for waste disposal, thus creating a potentially expensive waste disposal problem.[32] The accumulation of manure is an especially severe problem, "because so much is concentrated in so small an area at such a rapid, unending rate." [33] A dry-lot operation of 1,000 head of cattle has a sewage disposal problem equivalent to that of a city of 15,000 people.[34]

DRY-LOT FARMING IN THE EAST. Just as the intensive dry-lot system of livestock management need not be restricted to beef feeding, neither need it be restricted to the West, and it has actually begun to develop in the East. It may be expected to grow more rapidly as farms become larger and more specialized and the individual farm operator begins to concentrate on a single type of operation. Some farmers in the Corn Belt have already dispensed with livestock, and specialize in producing feed grains, which they sell for cash. Others have begun to concentrate on

[30] Examples of the manner in which a system of corrals may be laid out to facilitate the smooth and rapid movement of cattle to and from the milking barn are illustrated in Howard F. Gregor, "Industrialized Drylot Dairying: An Overview," *Economic Geography*, Vol. 39 (1963), 308–10.

[31] Gordon J. Fielding, "Dairying in Cities Designed to Keep People Out," *Professional Geographer*, Vol. 14, No. 1 (January 1962), 12–17.

[32] Feedlots clearly are for the birds; Timothy Lynch, Lloyd Tevis, and Rodolfo Ruibal, "Birds of a Cattle Feedlot in the Southern California Desert," *California Agriculture*, March 1973, pp. 4–6.

[33] Samuel A. Hart, "Manure Management," *California Agriculture*, December 1964, pp. 5–7.

[34] J. Van Dam and C. A. Perry, "Manure Management: Costs and Product Forms," *California Agriculture*, December 1968, pp. 12–13.

the feeding of livestock, whether cattle or hogs, and buy all of their feed. Perhaps the dry-lot system has developed most rapidly in the East along the fringes of Megalopolis, where dairy farmers feed their cattle citrus pulp from Florida, sugar beet pulp from Nebraska, and corn from Iowa, because it is cheaper for them to buy feed than to try to grow it. Throughout most of the East, however, one may confidently expect that the production of feed crops and the feeding of livestock, which traditionally have been integrated on each individual farm, will be conducted as separate operations on separate farms as farms become larger.

Farm Consolidation in Europe

Farms are being enlarged today in Europe, just as they are in the United States, but in Europe much of the process consists initially of consolidation, or rationalization of holdings, because many farms in Europe not only are woefully small by American standards, but also are fragmented into pathetically tiny parcels (Fig. 6–2).[35] In 1918, for example, the clachan of Rathlacklan in County Mayo, Ireland, had 56 families

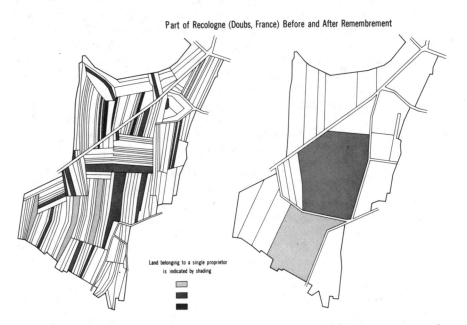

Part of Recologne (Doubs, France) Before and After Remembrement

Land belonging to a single proprietor
is indicated by shading

FIG. 6–2. *Consolidation of strips of farmland into solid blocks in part of the commune of Recologne (Doubs), France. Compiled from maps in Erich H. Jacoby,* Land Consolidation in Europe *(Wageningen, Netherlands: International Institute for Land Reclamation and Improvement, 1959).*

[35] Erich H. Jacoby, *Land Consolidation in Europe* (Wageningen, Netherlands: International Institute for Land Reclamation and Improvement, 1959); and Alan Mayhew, "Structural Reform and the Future of West German Agriculture," *Geographical Review*, Vol. 60 (1970), 54–68.

whose land was split into 1,500 fragments, some no more than a dozen square yards; one two-acre holding consisted of 18 separate plots. The redistribution and consolidation of holdings was not completed until 1942, and then "only at the cost of the physical collapse of one land-officer and the mental breakdown of another." [36]

Even today in parts of western Europe the pattern of landholding is best described by the graphic Italian term *polverizzazione.* "In the 1950s the average German farm comprised 20 plots totalling only 14 hectares (34.6 acres)," and in France "a typical mountain farm in Savoy, of barely 10 hectares (24.7 acres) consisted of 275 parcels; a village of 828 hectares (1,945.2 acres) in Loir-et-Cher had 5,075 plots, while extreme cases exist in which a holding consists of a few square meters, a grape vine, or a single olive tree. Parcels may lie 4 or 5 kilometers from the farm, and often a fifth of the working time is spent travelling between plots." [37] In part of Spain the average area belonging to each farmer was only 0.39 acres, and the average area per parcel was only 0.175 acres, but half the parcels contained less than 0.12 acres (a standard doubles tennis court covers 0.064 acres); grain was sown "on fragments of land small enough to be transported from one site to another by cart." [38] Farms which consist of small scattered parcels waste considerable potential farmland in such unproductive uses as access ways and boundaries; in the 1930s Long estimated that ten parishes in southwestern England had 1,641 miles of hedgerow, "about half again as long as the Great Wall of China." [39]

Excessive fragmentation of farmlands exists where the individual holdings are too small, or where the many tiny parcels comprising a single holding are too widely scattered; the two, unfortunately, often occur together. Many factors help to account for this fragmentation. The "fossilization" of the strip patterns of the old open-field system is an important one in many areas. In some places the land has been excessively subdivided under inheritance laws which require that an estate must be shared equally among all heirs. [40] In others the prestige value of land ownership is so great that farmers eagerly snap up any stray plot of land that comes onto the market, without regard to its distance from the farmstead or to the way in which it might fit into the farm operation. In alpine districts, and in specialized truck farming areas, where aspect or frost danger may be critical, the nature of the land or the type of farming may necessitate dispersal of farm plots as insurance against natural disaster.

[36] E. Estyn Evans, *Irish Folk Ways* (London: Routledge & Kegan Paul, 1957), p. 24.

[37] Audrey M. Lambert, "Farm Consolidation in Western Europe," *Geography,* Vol. 48 (1963), 31–38; this article has some striking illustrations.

[38] John Naylon, "Land Consolidation in Spain," *Annals,* Association of American Geographers, Vol. 49 (1959), 363.

[39] W. H. Long, "The Size of Fields in Devon," *Farm Economist,* Vol. 1 (1935), 224.

[40] Gottfried Pfeifer, "The Quality of Peasant Living in Central Europe," in William L. Thomas, ed., *Man's Role in Changing the Face of the Earth* (Chicago: University of Chicago Press, 1956), pp. 240–77.

Reorganization and rationalization of land holdings is a critical necessity in many parts of rural Western Europe today. The various schemes which have already been attempted range from the very simplest exchange of parcels between individual owners to ambitious programs for regional rationalization of holdings, but an enormous amount of work obviously remains to be done despite the progress which has already been made. And even after rationalization has been completed, many of the new holdings appear pitifully small when they are compared with contemporary American standards for similar types of farming. In Denmark, for example, where farm consolidation was completed in the middle of the nineteenth century, less than 10 percent of the land is in farms of 160 acres or larger, and in England it is estimated that few farms will ever exceed 500 acres.[41]

A primary explanation for this contrast in farm size is the basic difference in the approach to farmland in Europe, where land is dear and labor is cheap, and the United States, where land is cheap but labor is dear. This is another way of saying that the European farmer is prodigal of labor in attempting to wring out every last ounce of food that the land may be made to produce, whereas the American farmer uses his land prodigally (at least by European standards), but is vitally concerned to maximize his production per man-hour.

[41] Jacoby, *Land Consolidation in Europe,* footnote 35; and F. G. Sturrock and H. J. Gunn, "The Development of Large-Scale Farming," *Westminster Bank Review,* February 1968, pp. 59–65.

CHAPTER 7

factors that influence farmers' decisions

In order to understand any agricultural region, one must begin with a realization of the vital importance of the farmer as a decision-maker. Each individual farmer controls some segment of the earth's surface, and he alone makes the final decision as to how the land will be used, and how the operating unit will be organized for production of food and fiber. Any farm operation is a complex business, far more complex than the average city-dweller realizes, and in theory a farmer must integrate an enormous range of physical, economic, political, cultural, and perhaps other variables when he makes his decisions. In practice, much of this integration has already been performed many times over, and is available to the farmer in the tradition and experience of the community in which he farms, but the student of the rural landscape cannot afford to ignore the operation of variables which the farmer may be able to take for granted.

189948

Factors, Features, and Geographic Analysis

The variables which influence the farmer's decision might be thought of as the *factors* which create agricultural regions, and the existence of an agricultural region produces characteristic *features* upon the rural landscape. The distinction is seldom this neat and precise, however, because a landscape feature, such as a cotton gin or a sugar beet factory, which might appear to be the product of an agricultural system, may also be a powerful conservative force in maintaining its existence; to use a bit of jargon, the factors are the inputs of the system, and the features are the ouptputs, but feedback can convert outputs into inputs. Furthermore, the present state of knowledge is so limited that it is unwise to claim that the factors "create" the system, or that it "produces" features; it is much safer to think of both factors and features as the geographical covariants, or associates, of the agricultural region.

Although each of the factors which influences the farmer's decisions varies geographically, their geographical variations are not interrelated, and some vary much more rapidly than others. Variations in price support programs or length of growing season, for example, have little relationship to variations in soil drainage, or to each other; price support programs change abruptly at political boundaries, and the length of the growing season tends to vary only gradually over fairly large areas, whereas soil drainage conditions may vary considerably within a short distance on a single farm. For the most part, however, the factors which influence a farmer's decisions are essentially similar over fairly extensive areas, and the overwhelming majority of farmers make their decisions rationally in response to these factors. (Although economic determinists may prefer to ignore or overlook the fact, a decision based on tradition or on folk culture may be just as rational, as far as the decision-maker is concerned, as one based on the operation of the marketplace!)

Within a fairly extensive area of more-or-less similar conditions the majority of farmers, being rational men, tend to make more-or-less similar decisions, thereby creating an agricultural region. In a free society, an agricultural region might be defined as "an area in which large numbers of individual farm operators have made similar decisions in response to similar natural and human conditions, with similar effects." [1]

At one time, perhaps, a kind of "shotgun" approach to the geography of agricultural regions was justifiable, although such an approach fails to recognize the importance of the farmer as a decision-maker, and it tends to be disappointingly platitudinous and replete with obvious generalizations. It is not good enough, for example, merely to set down a long list of factors which purportedly "govern" crop production, apparently in the assumption that if enough items are listed, at least some of them are bound to be right.[2] Mere listing, unfortunately, is no substitute for analysis or understanding; it is a bit fatuous to list wind direction and precipitation as factors which are related to crop yields without examining the precise nature of the relationship.

Broad, bald generalizations are no longer necessary in geography, and they may not even be acceptable. Within the last decade or so scholars in the field have made increasing use of more rapid and precise quantitative tools—primarily mathematical and statistical techniques using electronic computers—for testing the strength of relationships, and these have permitted the examination and testing of a far greater range and variety of relationships than had hitherto been practical. At the same time, these tools necessitate a more rigorous scrutiny of the relationships

[1] John Fraser Hart, "Geographic Co-Variants of Types of Farming Areas," in E. S. Simpson, ed., *Agricultural Geography I. G. U. Symposium*, Research Paper No. 3 (Liverpool: University of Liverpool Department of Geography, 1965), p. 7.

[2] Lester E. Klimm, Otis P. Starkey, and Norman F. Hall, *Introductory Economic Geography*, 2nd ed. (New York: Harcourt Brace Jovanovich, 1940), pp. 127–29.

which deserve to be tested. A relationship for which no theoretical, no cause-and-effect, or no process justification can be postulated is one which presumably does not warrant testing, because there is no valid explanation for such a relationship in the embarrassing event that one should be detected.[3]

It is possible, for example, to demonstrate that spatial variations in the density of rural farm population on the Great Plains are directly correlated with spatial variations in average annual precipitation, but what is the theoretical explanation for this relationship?[4] Does anyone really believe that an increase in precipitation will grow more people, or that it will increase the fecundity of the people who are already on the ground? Perhaps precipitation and farm population density are actually the ends of a complex chain of relationships which might run something as follows: more rain brings better crop yields; better crop yields provide the same income on less land, which means smaller farm units; smaller farm units mean more people per square mile, assuming that the same number of people live on each farm unit, no matter what its size.

Each of the links in this chain appears to be a reasonable hypothesis which is theoretically justifiable, but the strength of each link should be carefully tested before making the quantum jump between the two ends. If the distant ends of the chain are indeed related, as they have been shown to be, then how much stronger should be the relationships between the individual links! And how much stronger geographers would be if, instead of examining relationships which have no theoretical basis, they would explore those relationships where some theoretical process of cause and effect may reasonably be postulated.

The first step in an approach toward the understanding of agricultural regions should be an attempt to isolate and analyze the individual components of farming systems. This would logically be followed by an examination of the manner in which each individual factor influences the farmer's decisions concerning that component, with a meaningful statement of the process or cause-and-effect relationship. The first step is feasible, and is attempted here, but unfortunately, given the present state of the art, the second is considerably more difficult, because some of the factors which influence the farmer's decisions have received more careful geographic attention than others. Much of what can be said must be more on the order of a blueprint for future research than a report on existing knowledge and understanding.

[3] The ready availability of packaged programs for factor analysis has generated an unconscionable number of "fishing trips" by geographers who apparently have had more computer time than ability to think through the theoretical significance of their variables; dumping in a lot of variables they have not understood has produced a lot of principal components and factor loadings and eigenvalues that they have not understood either, but the terminology is so impressive that the results simply "gotta be" scientific.

[4] Arthur H. Robinson, James B. Lindberg, and Leonard W. Brinkman, "A Correlation and Regression Analysis Applied to Rural Farm Population Densities in the Great Plains," *Annals*, Association of American Geographers, Vol. 51 (1961), 211–21.

Plants and the Habitat

The production of plants is an essential component of every agricultural system; plants can be produced without livestock, but every livestock operation depends upon plant production in one way or another, because plants are necessary for the transformation of mineral matter into nourishment for animals. Plant production, furthermore, is related to a wider range of factors than livestock production, because plants, unlike animals, are rooted in the environment; they cannot run, and they cannot hide. Animals can survive almost anywhere if suitable food and shelter are available, but plants are subject to every whim of nature. They must be able to withstand the bite of the wind, the scorching heat of the noonday sun, and the numbing cold of the long night watches. Their roots may be starved by a niggardly soil, alternately parched and waterlogged, or even laid bare completely by the ravages of erosion.

Cultivated plants, in short, are intimately related to the conditions of their habitat: precipitation, temperature, altitude, aspect, land form, and the nature of the soil. The relationships of plants with climatic and surface features have been so fully explored by agronomists and geographers that they need only be summarized here, despite their importance.[5] Any given area, and any farm within it, has available a climatically adapted ensemble of plants from which the farmer may select those which he wishes to grow; this ensemble, quite obviously, is far richer in some areas than it is in others. Nevertheless, no matter how rich or how poor it may be, the man who controls a given piece of ground is limited by climate in the kinds of plants he can grow on it, and this limitation will play a major role in determining the nature of his farm operation.

Climatic effects are regional, but those of landforms are local; within the constraints imposed by climate, the form of the land may impose further limitations, or it may offer certain possibilities. Altitude affects the climate; a rise in elevation tends to increase precipitation, but the temperature is depressed at an average rate of about 3.5° F per thousand feet. This effect is less important in areas of continental climate, where the annual march of temperature is rather abrupt, than in maritime areas, where the annual march is more gentle (Fig. 7–1). In the British Isles, for example, the upper limit of cultivation generally lies only 700 to 800 feet above sea level.[6]

The effects of slope are variable. In higher latitudes the slopes which face the sun receive its rays more perpendicularly than adjacent level areas, and enable farmers to grow more delicate plants, such as the vine. Slopes are also a favored location for frost-sensitive orchards, because they facilitate the drainage of cold air. In areas of heavy precipitation,

---

[5] Karl H. W. Klages, *Ecological Crop Geography* (New York: Macmillan, 1942) is a superb summary which deserves far wider recognition and use than geographers seem to have accorded it.

[6] John Fraser Hart, *The British Moorlands: A Problem in Land Utilization* (Athens, Ga.: University of Georgia Press, 1955), p. 15.

Diagrammatic comparison of mean monthly temperatures
(solid line), height above sea level, and length of growing
season in areas of maritime and continental climate.

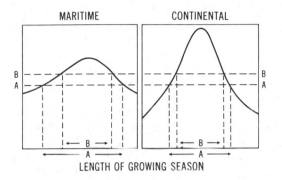

LENGTH OF GROWING SEASON

FIG. 7–1.   An identical increase in elevation shortens the growing season much more in areas of maritime climate than in areas of continental climate, because the annual march of temperature is much gentler in areas of maritime climate.

however, slopes are more susceptible to erosion, and they may require a protective vegetative cover of grass or even trees. In most hilly areas, where slopes are steep, the preferred cropland lies along the level valley bottoms, but bottomland may be liable to flooding, and like level areas elsewhere, it may require drainage. Despite the need for drainage, however, level lands have become increasingly popular as mechanization has increased and machines have grown larger, and in dry areas level lands have obvious advantages for irrigation.

The agricultural relationships of the soil are rather more awkward to deal with than those of climate and landforms, because the soil scientist prefers to tell the farmer what he *ought* to want to know about his soil, rather than what he actually does want to know. The farmer would like to know what he can do with his soil in terms of its fertility, workability, susceptibility to erosion, moisture retentiveness, and drainage requirements, whereas the soil scientist would rather discuss its origin, classification, shades of color, depths of horizons, structure, and other recondite technical details. From the farmer's point of view, a good soils map would summarize the conditions of climate, parent material, surface features, vegetation, and other physical factors which have a significant relationship to the agricultural decisions he must make, but unfortunately these are reflected only indirectly, if at all, in many soils maps, which are primarily concerned with soil origin.

The farmer is keenly aware, however, of the critical importance of maintaining the fertility of his soil, and this awareness influences many of the decisions he makes concerning his farm operation. Soil forms slowly, but under natural conditions the native plant life adjusts itself to the slow rate at which nutrients become available, and most of the nutrients removed from the soil are returned to it when dead plant tissues fall to the ground and decompose. The farmer upsets this balance when he replaces the native vegetation with crops which require greater quantities of nutrients, and he completely removes some of these nutrients from the farm when he sells them in the form of crops or livestock. Most farming systems return only a fraction of the nutrients removed from the soil, and the farmer must take positive action to replace them.

One of the most common modern techniques for replacing the nutrients removed from the soil is to supply them in the form of chemical fertilizers, although this is only one of a number of techniques which may be used to maintain soil fertility. One of the simplest is to let the land lie idle, or fallow, so that nutrients may have time to accumulate in it. Fallowing has been practiced since the very earliest days of farming, and it was the principal means of maintaining soil fertility in the three-field system of medieval Europe. Today fallowing is an essential ingredient of shifting cultivation systems in tropical areas, brush fallowing is still widely practiced in the southeastern United States, and a strip-crop system of fallowing is used in dry farming areas on the Great Plains.[7] Another technique for maintaining soil fertility is the use of green manure crops, which are not grown to be harvested, but solely to be plowed back into the soil to enhance its fertility. A green manure crop is especially useful if it is a legume, a plant whose root nodules accumulate the nitrogen which it is able to extract from the air.

Perhaps the most efficient way for a farmer to maintain the fertility of his soil is to ensure the return of the nutrients which his crops have removed from it, insofar as this is practicable. Normally only part of a plant is sold, and the stubble, stalks, stems, leaves, and other residues should be plowed back into the soil. The residues of crops which are consumed on the farm by livestock should also be returned to the soil in the form of barnyard manure, which increases its humus content and adds to its fertility.[8] Finally, the necessity of maintaining soil fertility is one of the principal reasons why a farmer should practice some kind of crop rotation.

Crop Rotations

Many farming regions have evolved a system of crop rotation which is in fairly close equilibrium with natural conditions, and has become more or less integrated with a certain kind of livestock operation. The integration of cropping system and livestock operation provides the foundation for a regionally dominant agricultural system which is modified

[7] J. E. Spencer, *Shifting Cultivation in Southeastern Asia,* Publications in Geography, Vol. 19 (Berkeley, Calif.: University of California Press, 1966); John Fraser Hart, *The Southeastern United States* (Princeton, N.J.: Van Nostrand Reinhold, 1967), p. 20; and F. J. Marschner, *Land Use and Its Patterns in the United States,* Agriculture Handbook No. 153 (Washington, D.C.: U.S. Department of Agriculture, 1959), p. 168.

[8] Eugene Mather and John Fraser Hart, "The Geography of Manure," *Land Economics,* Vol. 32 (1956), 25–38.

to fit the specific conditions and requirements of each individual farm. Crop rotations are such an essential element in good farm management systems that one is forced to wonder why students of agricultural geography have paid so little attention to the geography of crop rotation systems.

Crop rotations provide a far better model than crop combinations for understanding the cropping patterns of most areas. The notion of crop combinations is based upon the percentage of the total harvested cropland in a statistical unit, such as a county, which is in each crop.[9] The actual percentages are compared with a set of theoretical distributions in which equal percentages of the land are assigned to any given number of crops (100 percent if only one crop is grown, 50 percent each for two, 33.33 percent each for three, and so on to 10 percent for each crop if ten different crops are grown). Differences between the actual and theoretical percentages are calculated for the leading crop, the two leading crops, the three leading crops, and so on. These differences are squared, summated, and divided by the number of crops. The combination with the number of crops which yields the lowest value for a county is declared the crop combination for that particular county.

The concept of crop combinations is strictly mathematical. The lack of any explicit statement of the theoretical, process, or functional relationship upon which they are based leaves implicit the assumption that a farmer makes decisions concerning the crops he plants on a more or less random basis. The assumption that the crop he chooses to plant in one field, or in one year, is not influenced by functional relationships with those planted in other fields, or other years, might be unobjectionable in an area where a number of unrelated cash crops are produced, but it must be questioned in areas where crops are functionally integrated in a pattern of crop rotation.

Although a crop rotation system provides a model which can facilitate an understanding of the cropping system in the area where it is practiced, it is only a model, because any practical cropping system must be flexible. Few, if any, farmers adhere rigidly to one specific system, and even on a single farm the farmer may need to use different systems on different fields or different parts of the farm. Furthermore, he must be able to make year-to-year shifts in his cropping system in response to his expectations concerning costs of production, yields, and prices for alternative crops.[10] In addition, there is always the possibility that government programs might restrict the acreage of a crop which a farmer is permitted to plant in any given year.

[9] John C. Weaver, "Crop-Combination Regions in the Middle West," *Geographical Review*, Vol. 44 (1954), 175–200.

[10] In recent years some agricultural economists have had great fun showing how changes in the price of a given crop or crops would affect the acreage planted to it/them; one of the first studies was *Cotton: Supply, Demand, and Farm Resource Use*, Southern Cooperative Series Bulletin 110, published by the agricultural experiment stations of 13 southern states in cooperation with the Economic Research Service of the U.S. Department of Agriculture in 1966; see also Earl Swanson, "Price Expectations and 1969 Acreages of Corn and Soybeans in Illinois," *Illinois Research*, Spring 1969, pp. 3–4.

The farmer's options in modifying his cropping system are at least partially restricted, however, by his reluctance to disrupt his established crop and livestock patterns too severely. A farm is an integrated operation, and a drastic change in one phase might have serious consequences for another. The farmer must balance his expectations of short-term gains against his longer-range requirements. He might well be foolhardly, for example, to cash in on the accumulated fertility of his soil simply because he expects the price for a given crop to be unusually high in one particular year, and he would be understandably reluctant to grow two crops whose seasonal labor peaks overlapped too closely, or to change to a cropping system which failed to provide enough feed for an established livestock operation.

The livestock operation which has been developed on many farms is closely interrelated with the crop rotation. The crops produced on a farm may be divided into food, cash, and feed crops; some may provide food for the farmer and his family, and some may be sold for cash, but most rotation systems produce some crops which are used most efficiently if they are fed to livestock. The crop rotation does not determine the livestock system; the same crops can support quite different livestock systems, identical livestock systems can be supported by a variety of crops, and some farmers depend largely or entirely upon feed purchased from other farmers. The livestock potential of many farms, however, depends in large measure on the products made available by the cropping system, and the livestock system, in return, helps to maintain the fertility of the land from which the crops are produced.

The crop rotation system, in short, provides a fundamental point of departure for understanding an entire farming system. A good rotation is nicely adjusted to the physical conditions of the land. It produces crops which support the livestock operation. The livestock operation helps to maintain the land in good condition for growing crops. A successful farming system, which integrates a good crop rotation with a profitable livestock operation, tends to become dominant, with appropriate modifications, over an extensive area. Substantial variations in farming systems, whether in time or in space, are easier to comprehend if one understands the basic system and the crop rotation upon which it is founded.

The Selection of Livestock

The livestock system of a farm or region is related to the kind and amount of feed which is available for animals. Under the Norfolk four-course rotation (wheat, roots, barley, ley) in Britain, for example, the two cash crops, wheat and barley, also provided straw, which served both as roughage for the cattle and as litter to catch their droppings. At first the principal root crops were fodder roots (turnips, swedes, or mangolds), which provided winter feed for cattle and sheep, but in some areas these were eventually replaced by sugar beets or potatoes, which were sold as cash crops. Sugar beet tops and dried pulp have about half the feeding value of fodder roots, and thus could continue to support a fair number of sheep and cattle, but the farmer who switched to potatoes was forced to reduce the size of his livestock operation considerably for lack of feed. Part of the ley was set aside and cut for hay, and the rest was grazed.

The Norfolk rotation supported both sheep and cattle. A flock of ewes, which produced both wool and lambs, was pastured on the grass fields in summer, and on the roots in winter. The sheep were enclosed in temporary folds made of heavy wicker hurdles, which could be moved around the field to ensure that both grazing and droppings were concentrated. The sheep remained on the farm throughout the year, but cattle were bought lean in the fall to be fattened during the winter and spring in large enclosures, or cattle yards, behind the barn. They were fed roots, straw from the grain which was threshed in the barn, and a rich concentrate compounded of cottonseed and linseed. Although fat cattle were sold, a primary purpose of the cattle-feeding operation was to produce manure, which was spread on the old grassland before it was plowed up for wheat. The Norfolk system enriched the soil with sheep droppings and cattle manure, and it enriched the farmer by producing wheat, barley, lambs, wool, and cattle which he could sell.

Although the Norfolk system still provides a useful model for understanding British farming, the system itself has been modified extensively in the years since World War I. One of its weakest links was the fodder root crop, which demanded considerable labor but did not provide an especially good livestock feed. The death knell of the sheep flock was sounded early on by the enormous amount of labor required for folding sheep on roots. Beef could be imported more cheaply than it could be produced at home, which encouraged many farmers to switch to dairy farming, with greater emphasis on grass. Fodder roots have been replaced on many farms by cash crops of sugar beets and potatoes. The development and increased use of chemical weedkillers and fertilizers have reduced the need for row crops and barnyard manure, and grain crops can be produced on a much larger scale with new and improved machinery. Much of the grain is fed to dairy cattle and hogs.

THE MIDDLE WEST. The traditional three-year crop rotation in the American Middle West has been corn in the first year, a small grain (wheat or oats) in the second, and a hay crop in the third. Most of the corn crop is harvested for grain, and most of the grain is fed to livestock. Approximately five of every ten ears of corn harvested are fed to hogs, three are fed to cattle, one is fed to other kinds of livestock, and the tenth provides the raw material for industries manufacturing some 600 different products, from breakfast cereals to whiskey. Most of the wheat is sold as a cash crop, but the oats are more likely to be fed to livestock on the farm. Oats are not as important as they were in the horse and buggy days, when they were fed to horses, but they still are an excellent feed for young animals and breeding stock.

How does a Midwestern farmer choose between hogs and cattle, or between beef cattle and dairy cattle? Cattle are more prestigious, demand less work, and many hog farmers dream of the day when they can "trade up" from hogs to cattle. Hogs are better than cattle at converting corn and other concentrated feeds into meat, but cattle are more efficient converters of roughages such as silage, hay, and pasture. Farmers with plenty of hay and pasture land lean toward cattle, but the farmer who produces quantities of concentrated feeds does not always turn to hogs. Hogs and corn compete for the same labor, and the size of the hog operation tends to decrease as the amount of corn produced on the farm increases. A hog farmer breeds his own animals, because hogs eat like hogs, gain better than a pound a day, and reach a market weight of 200 pounds in six months. A hog farmer must tend to his hogs at the same time that he is busy with his crops, but a beef farmer can afford to pay a rancher to put the bone and hide on lean feeder calves and yearlings; he buys them in the fall after he has harvested the crops on which he will fatten them in his feedlot.

As a general rule, the farmer with a small corn acreage can afford the time to fatten hogs, the intermediate farmer fattens cattle, and the farmer with a large corn acreage does not have time to raise livestock. Within the Corn Belt, therefore, the small farms of Ohio and Indiana tend to concentrate on hogs, the medium-sized farms of Iowa tend to concentrate on fattening beef cattle which have been shipped in from ranch areas to the west, and the large farms of central Illinois sell their corn and soybeans as cash grain rather than feeding them to livestock. Farmers in the Dairy Belt to the north produce less concentrated feed and a greater abundance of roughages, such as corn silage, hay, and pasture, which are better suited to the production of milk rather than meat; who ever heard of silage-fed beef?

Other Factors

The physical environmental factors which influence a farmer's decisions are most directly related to his selection of plants and crop rotations, whereas his decisions concerning his livestock operation, as well as many of his decisions concerning crops, are influenced by a far wider range of factors. Among the most important of these is the balance between costs of production and prices of farm products, which largely determines the income of the farmer. The price mechanism, in theory, represents an integration of all factors which influence the farmer's decisions, and on the gross scale this is probably true, but perhaps it does not operate quite so neatly at the level of the individual farmer. Most of us do things from time to time that do not make sense in terms of dollars; why should we expect farmers to be any different?

Most students of the geography of agriculture have focused on the subject from the point of view of economics, which is obviously important.[11] No one can argue that the balance of costs and prices will

---

[11] It seems unnecessary to repeat here material which has already appeared in Leslie Symons, *Agricultural Geography* (London: G. Bell, 1967), and Michael Chisholm, "Problems in the Classification and Use of Farming-Type Regions," *Transactions and Papers*, Institute of British Geographers, No. 35 (1964), 91–103.

not, for the most part, determine which commodities out of the physically practical range will be produced on a given farm, but I prefer to examine some of the factors which have received rather less attention (although they may well be reflected in the price mechanism) despite the fact that they most certainly influence the decisions of farmers and the look of the rural landscape.

FARM SIZE AND FARM TENURE. The relationships between the decisions made by farmers and farm size, farm enlargement, farm tenure, and farm management systems have already been discussed in considerable detail. Other things being equal, a small farm requires more intensive application of labor than a large one in order to provide an acceptable level of living. The man who has a small farm generally tries to produce a more valuable farm product, or a greater quantity, or both, from a given acreage of land. In the Middle West, for example, one farmer might choose to fatten beef cattle because his crop work did not leave him enough time for hogs, whereas his neighbor on a smaller farm might concentrate on hog production because this enabled him to make effective use of the time he had available.

A tenant farmer, unless he has considerable security of tenure, is reluctant to make any substantial investment in the land or buildings of the farm unless he is fairly sure that he can take it with him when he leaves. A tenant farmer, for example, is not likely to plant an orchard or vineyard, and he is probably more interested in cash crops, which will yield a quick return, than in trying to build up a herd of livestock, which might take years. The tenant farmer is often accused of mining the fertility of the soil, and he is notorious for his lack of interest in the appearance and upkeep of the farmhouse and other farm buildings.

PHYSICAL DEMAND. The physical (as distinct from economic) demand for certain specialized farm products has a powerful influence on the location in which, and the acreage upon which, farmers are able to conduct some kinds of agricultural activities. For example, such perishable commodities as milk, fruits, vegetables, and perhaps even meat in some instances, must be produced close to the people who will consume them, or close to transportation facilities which can deliver them to the consumers quickly and efficiently. For this reason dairy farms, truck farms, and greenhouse districts are near many urban centers.

The physical inability of people to consume more than a limited quantity of certain farm products, such as potatoes, peanuts, and tobacco, restricts the amount of land on which farmers can grow such crops. In the United States between 1961 and 1965, for example, tobacco was grown on an average of only about 1.17 million acres, which was roughly one twentieth of 1 percent of the land area of the nation, or one of every 2,000 acres. Tobacco easily could have been grown on a much larger area, as far as physical conditions are concerned, but this limited acreage produced 1.86 billion pounds, which would have accounted for slightly more than the total amount of tobacco actually consumed in the United States if it had all been made into cigarettes, or five-sixths of the total national consumption of tobacco.

The physical demand for certain kinds of crops and animals is closely related to the dietary habits of people, the ideas they have about what is fit to eat and what is not, and the foods they prefer and the foods they avoid. Although no one, to the best of my knowledge, has studied the geography of dietary habits in the United States very carefully, most of us would have a pretty fair idea of what part of the country we were in if we were offered gefilte fish, or gjetost, or grits, or guacamole. Eggs, for most of us, come from chickens, and not from geese, or turkeys, or ducks, or pheasants, or ostriches, and milk comes from cows, not mares, or sheep, or goats, or sows, or whales. In 1965 Japanese farmers produced one thirty-third as much wheat, one-third as many potatoes, and more than 4.5 times as much rice as farmers in the United States. The demand for dog meat is not what it used to be back in the good old days, when Indians roamed the North American continent, and fattened dogs especially for their feast days; it was a signal honor for a guest to be served the head of a plump roast dog.

Dogs and horses are much more fastidious eaters than hogs, which happily grow fat on garbage and slop, but most Americans, who like to "eat high on the hog," have never knowingly tasted horse meat, and are slightly revolted at the very thought of eating dog. Even mutton and lamb, which are quite popular in some parts of the world, are not commonly eaten in the United States. Why do Americans eat more than 17 times as much pork as lamb and mutton? One possible line of explanation, which has not been carefully explored, would begin with the suggestion that most Americans have never had a chance to taste really good mutton or lamb, because farmers in the Corn Belt, where the nation's finest meat is produced, are strongly prejudiced against sheep and refuse to have them on their farms. Could this prejudice be traced back to the sandy lands of the North European Plain, where sheep were a menace because they grazed the heathland vegetation so closely that the sandy soil was exposed to wind erosion and destroyed? Such an explantion would seem to make at least as much sense as more commonly proposed explanations based on wild dogs, fencing problems, low prices, or diseases.

PRICE-SUPPORT PROGRAMS. Most politicians are very much aware of the fact that farmers, like the rest of us, have the right to vote, and so the politicians have passed laws which they think will help farmers, or at least, which will get their votes. Unfortunately, politicians seem to have a knack for causing trouble when they start meddling with agriculture, and many farm programs do not appear to work very well. Nearly every country in the world, for example, has some kind of farm price-support program, which almost inevitably becomes all tangled up with marketing arrangements, acreage and production controls, tariffs and import restrictions, export subsidies, and a host of official regulations.

The farm price-support program in the United States, which dates from the depression years of the 1920s and 30s, has been quite simple in theory: farmers cannot afford to withhold their crops from the market, even temporarily; supply exceeds demand when all farmers sell their crops at the same time, and the price falls temporarily, but often severely; the government should help farmers to withhold this temporary surplus until demand revives by giving them loans equal to the amount they would receive if they could sell their crops at a fair and reasonable price. In practice, of course, the loan price became a support price for the crop, because any farmer would have been foolish to have sold his crop to any buyer who offered less. Each loan was made for a year; if the market price failed to rise above the loan price in that year, the farmer let the government keep his crop, and thus it became part of the infamous "government surplus." [12]

A guaranteed price for any crop would encourage many farmers to start growing it, which would aggravate the surplus problem, and therefore they have had to be discouraged by a system of controls known as "acreage allotments." If the surplus of any crop becomes too great, the Secretary of Agriculture is required by law to reduce the acreage on which the price will be supported in the following year. This reduced acreage is "allotted" to individual states on the basis of the acreage of that crop which their farmers planted in a selected base year; within the states this reduced acreage is allotted to counties, and within counties, to individual farmers, so that each farmer who grew the crop in the base year receives an "allotment," which is the acreage of that crop he is permitted to plant. He is severely penalized if he grows more than his allotted acreage, but he receives a guaranteed price for all of the crop he can produce on the acreage he has been allotted.

In order to maintain his income from reduced acreage, the clever farmer plants his allotted acreage on his best land, and pours on the fertilizer to increase his yield, thus setting in motion a vicious cycle. [13] Increased yields from a reduced acreage continue to produce a surplus, which requires another acreage reduction and a further decrease in the size of allotments. [14] Once again the farmer attempts to increase his yields, and the cycle is broken only when bad weather strikes the crop, or when a war comes along to absorb the surplus.

Furthermore, a guaranteed price in a major producing nation tends to set a floor for the world price, and farmers in other countries are encouraged to start growing the crop. The home market must be protected from their competition by tariffs and import restrictions, and exports of the crop must be subsidized to enable them to compete on the world market. Perhaps the ultimate absurdity has been reached in cotton: cotton textile manufacturers in foreign countries can buy American cotton at a

---

[12] Loans were replaced by direct subsidy payments in the farm bill which was signed into law in August 1973. The new farm program is based on the key concept of a "target price" for wheat, feed grains, and cotton. The government will pay farmers a direct subsidy to make up the shortage whenever the free market price for a crop drops below its target price, but the Secretary of Agriculture is empowered to restrict the acreage of any crop if the subsidies on it become excessive.

[13] P. Weisgerber, *Productivity of Diverted Cropland,* Economic Research Service 398 (Washington, D.C.: U.S. Department of Agriculture, 1969).

[14] Raymond P. Christensen and Ronald O. Aines, *Economic Effects of Acreage Control Programs in the 1950's,* Agricultural Economic Report No. 18 (Washington, D.C.: U.S. Department of Agriculture, 1962).

lower price than domestic manufacturers, because cotton exports are sub-sidized, and the textiles they make from this cheaper cotton can under-sell domestic textiles on the domestic market.

Farm price-support and acreage-allotment programs have tended to fossilize the geography of crop production by keeping individual crops concentrated in the areas where they were produced in the base year. An acreage allotment is a share in a monopoly, established and guaranteed by the government, for the production of a given crop, and a farmer would be foolish if he failed to hang on to his allotment even when he cannot use it very effectively. Furthermore, price-support programs have given thousands of farmers just enough income to enable them to cling to undersized farms which do not really provide a satisfactory level of living, while the lion's share of the money spent on support programs has gone to large, efficient farms which do not need it. The billions of dol-lars devoted to price-support programs each year might have been better spent if they had been used in programs to help farm families escape their undersized farms and make better lives for themselves elsewhere.

GOVERNMENT ASSISTANCE PROGRAMS.   In addition to price-support programs, many other government programs have helped to influence the decisions of farmers and change the look of the land. The construction of dams and ditches has extended irrigation in the West, and grants toward the costs of drainage works have facilitated reclamation and use of wet-lands in the East. Materials, services, and financial aid have been provided for a variety of soil conservation activities, such as contour plowing, strip cropping, construction of farm ponds and terraces, and diversion of crop-land.[15] Perhaps most important of all has been an extensive federal pro-gram of education, experimentation, technical assistance, and informa-tion dissemination, which has been conducted in cooperation with the land grant colleges and universities. The agricultural experiment stations in each state publish technical and popular bulletins of great value, and the extension agents in each county can be a mine of information both for farmers and for students of farming.

The tremendous potential impact of government assistance programs is demonstrated by contrasts in the advance and retreat of settlement on either side of the provincial boundary which bisects the Great Clay Belt in Canada.[16] Between 1931 and 1957 land was abandoned and settle-ment retreated on the Ontario side of this boundary, where the provincial government had essentially no program of assistance. On the Quebec side, however, the settled area was doubled by an official program of colonization which included presettlement soil surveys; a policy of settle-ment in compact blocks; careful screening of prospective settlers; finan-

15 The Agricultural Stabilization and Conservation Service of the U.S. Depart-ment of Agriculture reports on these activities each year in publications entitled *Agricultural Conservation Program: Maps* and *Conservation Reserve Program: Statis-tical Summary.*

16 George L. McDermott, "Frontiers of Settlement in the Great Clay Belt, Ontario and Quebec," *Annals*, Association of American Geographers, Vol. 51 (1961), 261–73.

cial assistance for temporary living expenses, land clearing, plowing, construction of houses and farm buildings, and purchase of livestock and farm equipment; provision of heavy machinery for bulldozing stumps, deep plowing, and digging drainage ditches; and the availability of expert technical assistance and advice.

KNOW-HOW AND FACILITIES. An astonishingly large number of city people today seem to think that almost anybody can be a farmer, whereas in actual fact the modern farmer is more at home in the city than a city person is in the countryside. Most city dwellers, if they are really honest about it, would have to admit that they do not know how to dehorn cattle, or shear sheep, or castrate hogs, nor would they know when to plant wheat, or sucker tobacco, or pick corn, nor could they adjust and operate a hay baler, or a combine harvester, or a milking machine, although it would be difficult to make a living on almost any farm in the United States without possessing at least some of these skills.

Any farming system, to be successful, requires managers and workers (many farmers wear both hats simultaneously) who have the necessary skills and knowledge, and it also requires the availability of appropriate facilities and equipment for production and marketing. In most established farming regions these skills and facilities have developed through time. Farm people have the knowledge, experience, and tradition of growing the necessary crops, from preparation of the soil to harvesting and processing, and of taking care of the necessary kinds of livestock. Most farms have grown to about the right size, and each farm has the buildings and machinery needed to raise the traditional crops and livestock. In many areas the basic crops are protected by acreage allotments and price-support programs. The local road system is good enough to get farmers to a service center which has knowledgeable merchants with the facilities necessary to buy and/or process their products, and bankers who understand their credit needs.

In most successful farming areas the local infrastructure of skills and facilities can be taken for granted, and once established, it plays a powerful conservative role in influencing the decisions of farmers and maintaining the traditional farming system.[17] It has evolved from the experiences of many farmers over a long period of time, and represents an integration of many of the variables which should influence the decisions of farmers in the area. It largely relieves the individual farmer, as he makes his decisions, of the necessity of having to consider each one of these variables, because he can base his decisions upon their summation, as embodied in local experience and tradition, and the existing infrastructure of skills and facilities. A farmer may be able to take many of these variables for granted, or even to ignore them completely, but the student of the rural landscape cannot afford to do so.

[17] For a discussion of the role of the cotton gin in fossilizing the geography of cotton production, see Merle C. Prunty and Charles S. Aiken, "The Demise of the Piedmont Cotton Region," *Annals*, Association of American Geographers, Vol. 62 (1972), 283–306.

The vital importance of infrastructural facilities and skills becomes dramatically apparent when the farmers of an area desire, or are compelled, to change to a new farming system. Farmers and farm workers must learn to cope with unfamiliar crops and different kinds of livestock, and their ignorance may result in heavy losses, at least in the beginning. The new farming system may require large cash outlays for the purchase or construction of equipment and facilities which had not hitherto been necessary, plus the scrapping of existing ones, which may well represent a considerable investment of capital. Marketing and credit arrangements, which can be taken for granted in traditional farming areas, may have to be developed from scratch.

The changeover from cotton to cattle in parts of the South since the Second World War highlights the kinds of stresses and strains which can be produced by attempts to change a farming system. The farmhand's cotton-picking skills did not help him very much when it came to taking care of livestock, and nothing in the planter's experience helped him know where to go to buy cattle, or how to judge their quality when he finally got there. The old cotton fields had to be converted into pastures, and the planter discovered that some infernal weeds had suddenly become wonderful forage crops. The planter had never had to worry about his cotton getting out onto the highway and being run over by a truck, but the cattle on his new pastures had to be enclosed by fences, and provided with drinking water. All these things cost money, but when he tried to borrow it the planter found that he had to educate his banker to the fact that the gestation period for cows is a bit longer than the six months of a traditional bank loan on cotton. And finally, when the cattle were ready for market at last, the planter found that there was no good place to sell them; I knew one Georgia cattleman who became so exasperated with the local marketing situation that he trucked his cattle all the way to St. Louis, a mileage greater than the distance which separates Omaha and Chicago, or St. Paul and Kansas City.

With all the problems which face them if they attempt to make any great change, it is small wonder that most farmers are conservative businessmen, and quite wary of change. Apart from the pressure of necessity, how does an agricultural region ever change? Much of the answer appears to lie in the activities of enlightened innovators, whose successful ideas are eventually emulated by their neighbors. And high on the list of innovators is that much maligned individual, the gentleman farmer, to whom most farmers owe a far greater debt than is ever acknowledged. The man who is interested in, and can afford to tinker with, new ideas has been responsible for many, if not most, of the major improvements which have been made in agriculture, whether one considers the Enclosure Movement, the Agricultural Revolution, or the last third of the Twentieth Century.

CHAPTER 8 *farm buildings*

Each farmer makes his own decisions concerning the kinds of crops he will grow and the kinds of livestock he will raise. When thousands of farmers over a large area make similar decisions, they create an agricultural region, where essentially the same kinds of crops and livestock are produced. One might say, as a working model, that each agricultural region has its own distinctive rural landscape, because: (a) the lie of the land is such an important factor influencing farmers' decisions that any agricultural region tends to have similar surface features; (b) similar decisions concerning the plants which will be grown throughout an agricultural region give it a more or less homogeneous vegetative cover; and (c) the successful production of each kind of crop and each kind of livestock requires its own more or less distinctive structure or structures, and the farm operation which produces a given combination of crops and livestock requires a distinctive ensemble of such structures; in other words, each agricultural region has its own characteristic type of farmstead.

Farmsteads must be analyzed in terms of the individual structures comprising them, which in turn must be seen in terms of their relationship to the crops or animals requiring such structures. On most farms the farmstead is shaded by trees, and in the drier western grasslands many farmsteads have protective windbreaks planted along their northern and western sides.[1] The most essential structure of the farmstead (if it has indoor plumbing) is the farmhouse which shelters the farmer and his family. Behind the farmhouse are the shelters and handling facilities for livestock, and structures in which crops are stored and processed; the structures associated with crop production are more likely to be in the

[1] *Planning Your Farmstead: Arrangement, Buildings, Landscaping, Windbreak,* Circular 732 (Urbana, Ill.: Illinois Agricultural Experiment Station, 1960).

115

fields where the crops are grown.[2] Most processing and marketing structures for crops and livestock are in the central place which serves the farm. The farmstead must have protective storage for implements, machinery, fuel, and fertilizer, and most modern farms need a workshop which is well equipped with tools and spare parts.

Some of the buildings of the farmstead are multipurpose structures which are put to a variety of uses at different times of the year, but others, such as the oast house or the mint still, serve only a single specific function. The appearance of some structures, such as tobacco barns, is largely determined by the function they are intended to perform, but the farmer has considerably more leeway in designing others, such as fences, and their appearance may vary considerably from region to region, or from one culture group to another within a region.[3] Some structures, however, such as garages, workshops, and privies, do not seem to have offered any great challenge to the decorative ingenuity of farmers and their wives, because they have fairly much the same nondescript appearance in all parts of the country and among all culture groups.

Individual Structures

LIVESTOCK SHELTER AND HANDLING.   The farmer who has livestock must protect them from wind, snow, cold rain, excessive sunshine, and extremes of temperature. The structure which houses the animals should permit the farmer to feed and handle them with convenience and safety, both to man and to beast. It should have an adequate water supply, and it should permit easy cleaning and removal of manure. On many small farms in the old days the horses, cattle, hogs, and other livestock were all kept in a single general purpose barn, which was also used to store grain and hay. Most of the larger farms had a separate horse barn for the work stock, but the replacement of horses by tractors has made the old horse barn unnecessary, just as increasing specialization has made general purpose barns obsolete, and both types have been converted to other uses, torn down, or allowed to fall down.

Dairy cattle traditionally have been housed in barns with large hay mows on the upper level; the ground level had individual milking stalls

[2] Wayne E. Kiefer, "An Agricultural Settlement Complex in Indiana," *Annals*, Association of American Geographers, Vol. 62 (1972), 487–506. A superb book, with which anyone interested in the country side ought to be familiar, is Amos Long, Jr., *The Pennsylvania German Family Farm*, Publications of The Pennsylvania German Society, Volume VI (Breinigsville, Pa.: Pennsylvania German Society, 1972); anyone interested in southeastern Pennsylvania should rush right out and buy a copy immediately.

[3] John Fraser Hart and Eugene Cotton Mather, "The Character of Tobacco Barns and Their Role in the Tobacco Economy of the United States," *Annals*, Association of American Geographers, Vol. 51 (1961), 274–93; Fédération Nationale des Planteurs de Tabac en France, *Dessiccation et Séchoirs à Tabac en France* (Paris: Editions S. E. D. A., 15 rue du Louvre, n.d.); and Eugene Cotton Mather and John Fraser Hart, "Fences and Farms," *Geographical Review*, Vol. 44 (1954), 201–23.

at which the cows were kept tied up in cold weather, except for a short exercise period each day. In recent years, however, some dairy farmers, especially those in milder climates, have begun to use a loose housing system; the cows are confined in stalls only to be milked, and spend the rest of their time in a yard with an open shelter along one side.[4] The cows are milked, a few at a time, in a milking room located in a small building next to the yard. With either system, the dairy farmer must have a separate milkroom, where milk can be cooled and stored until it is picked up. State and local health regulations closely control the design, construction, and sanitation of farm dairy buildings, and the increasing stringency of these regulations has squeezed out of the dairy business many small farmers who have been unable or unwilling to make expensive modifications of their existing buildings.

Hogs, sheep, poultry, and other animals are housed in small buildings scattered around the farmstead. Hogs may be kept in a permanent structure located near the other farm buildings, but some hog farmers prefer individual hog houses which can be moved around the farm from field to field and always kept on clean ground, thus reducing the risk of disease, to which hogs seem especially susceptible.[5] A permanent hog house, with separate pens for each sow, might be 20 to 30 feet wide and 30 to 40 feet long, with three or four low hog doors two feet wide and three feet high on each side. A movable individual hog house seldom measures much more than six by eight feet, so that the sow's body heat can keep it at a comfortable temperature in winter.[6]

Many farms have small poultry houses in which hens used to be kept to provide eggs and meat for the family table, and "egg money" for the farm wife. These houses are rarely more than eight to ten feet high, to reduce air space and heating requirements; the west, north, and east sides are solid, but the south side is almost all windows, to admit heat and light. The farm poultry house is seldom used for chickens any more; a few have been converted into hog houses, but the majority stand empty, or are accumulating junk. Commercial poultry houses, which are appreciably larger, are used to produce either broilers or eggs. Broiler houses are long, low, one-story buildings which can house 10,000 to 15,000 birds at a time.[7] An unusual kind of relict poultry house is the enormous dovecotes which are still to be seen in parts of England.[8] During the late Middle Ages pigeons from the dovecote were a delicacy

[4] *Loose-Housing System for Dairy Cattle*, Miscellaneous Publication No. 859 (Washington, D.C.: U.S. Department of Agriculture, 1961).
[5] T. E. Bond and G. M. Petersen, *Hog Houses*, Miscellaneous Publication No. 744 (Washington, D.C.: U.S. Department of Agriculture, 1958).
[6] Kiefer, "An Agricultural Settlement Complex," footnote 2, p. 505.
[7] John Fraser Hart, *The Southeastern United States* (Princeton, N.J.: Van Nostrand Reinhold, 1967), pp. 40–43. The evolution of styles of poultry houses which are now relict features in the former "Egg Basket of the World" are described in John Passerello, "Adaptation of House Type to Changing Function: A Sequence of Chicken House Styles in Petaluma," *California Geographer*, Vol. 5 (1964), 69–74.
[8] Olive Cook and Edwin Smith, *English Cottages and Farmhouses* (London: Studio Publications, 1955), pp. 220–26.

in late winter, when other fresh meat was scarce, and their droppings were valued as fertilizer.[9]

Most forms of livestock require enclosures and handling facilities as well as shelters, and nearly every farm has some kind of collecting and holding area, which may be called a barnyard, feedlot, corral, paddock, or something else, depending on the farm and the kind of livestock. Close at hand one might find weighing scales; loading ramps; and chutes or squeeze gates, where animals can be held fast for dehorning, trimming, and other veterinary work. The barnyard must have fences stout enough to hold the animals of the farm, and the lanes connecting the farmstead with fields and pastures must be lined with fences. The pastures must be fenced to keep the animals in them, and the fields must be fenced to keep animals out.[10]

CROP PRODUCTION. The fences which enclose fields and pastures are only one of several types of features associated with crop production which are located in the fields rather than at the farmstead. Orchards and vineyards are an obvious example, as are the greenhouses of truck farming districts and urban fringe areas. The layout of the fields themselves may be closely related to the farming system. A striking illustration of such a relationship is the alternation of strips of wheat and fallow land to conserve soil moisture in dry farming areas.[11] Other soil conservation practices, such as contour cultivation, and the construction of low terraces to reduce runoff, also leave their distinctive marks upon the landscape. Low earthen terraces, which snake through field, forest, and pasture alike in many parts of the modern South, are especially useful relict features for the student of land use. They indicate the former extent of cultivated land, because they were originally constructed to reduce the danger of erosion in cotton fields.

Crop production in much of the West is not possible without irrigation systems, which can produce distinctive features on the landscape. Most major irrigation projects require massive dams and large storage reservoirs, or deep wells and powerful pumps, with a close network of canals and ditches to get water to the fields. Center pivot sprinkler irrigation systems, which came into general use in the 1960s, produce circular "fields" which stand out quite strikingly on the rectilinear American landscape. Paradoxically, areas which have been irrigated also require carefully planned drainage systems to carry away alkaline ground water and to prevent the soil from becoming waterlogged.

Large areas in the humid East, including some of the most productive parts of the Corn Belt, could not have been cultivated without the development of extensive drainage works. Geographers have largely ignored the importance of drainage, perhaps because its end product is

[9] M. S. Seebohm, *The Evolution of the English Farm*, rev. 2nd ed. (London: Allen and Unwin, 1952), pp. 136 and 211.

[10] Mather and Hart, "Fences and Farms," footnote 3.

[11] *The Look of Our Land, An Airphoto Atlas of the Rural United States: The Plains and Prairies*, Agriculture Handbook No. 419 (Washington, D.C.: U.S. Department of Agriculture, 1971), pp. 4–5.

similarity to adjacent areas, rather than distinctiveness from them.[12] Apart from large open ditches, drainage works are not readily visible, although the herringbone pattern of buried tile drains sometimes is strikingly apparent from the air.

Some features associated with the harvesting and temporary storage of crops are also in the fields rather than at the farmstead. On cotton plantations in the South a portable cotton shed was placed beside each sharecropper's plot at harvest time, and he stored his cotton in it until he had picked enough to gin out a bale.[13] After peanut vines have been lifted, they are stacked around poles in the field to dry for a few weeks before the nuts are removed. Farmers in parts of Appalachia still cut stalks of corn when they are ripe, stack them in shocks in the field, and shuck out the ears as they are needed during the winter. Many European farmers, after they have harvested their heavy fodder roots, store them in pits, or "root clamps," dug in the field where the roots were grown, which saves the labor of hauling them back to the farmstead.[14] And the field structures associated with the hay harvest are so enormously varied that they merit separate treatment.

HAY AND FORAGE CROPS. Grasses and legumes are grown more widely, and on a larger acreage, than any other crop. The field on which they are grown may be used as a pasture, if it is suitably fenced, and the crop may be harvested by grazing livestock, but much of it is mowed and stored as hay or silage for winter feed. In the United States swathes of hay, which may be turned occasionally, are left on the ground to dry, but in the milder, moister parts of Europe farmers have learned from sad experience that they cannot count on the sun to cure the hay crop unless they keep it off the damp ground, and they have developed a variety of ingenious devices for doing so. In Alpine areas the branches are lopped off young trees fairly close to the trunk, the trunk is then driven into the ground, and hay is hung on the branches. From a distance these "little hay men" look almost like people standing in the hay fields, whence their colloquial name.

Many farmers in Europe stack their hay on wooden tripods to keep it off the ground. In parts of Denmark the tripod is replaced by a quadripod, which permits air to flow beneath the drying crop, as well as around it; quadripods are especially useful for drying the tops of sugar beets. Farther north, in the cooler and damper areas of Norway and Sweden, hay is hung on lines strung between posts which have been driven into the ground, very much as a housewife hangs her weekly washing on the family clothesline. In Slovenia hay is hung on horizontal wooden cross-

[12] Leslie Hewes and Phillip E. Frandson, "Occupying the Wet Prairie: The Role of Artificial Drainage in Story County, Iowa," *Annals,* Association of American Geographers, Vol. 42 (1952), 24–50.

[13] Merle Prunty, Jr., "The Renaissance of the Southern Plantation," *Geographical Review,* Vol. 45 (1955), 473 and 476.

[14] John Fraser Hart, "Vestergaard: A Farm in Denmark," in Richard S. Thoman and Donald J. Patton, eds., *Focus on Geographic Activity: A Collection of Original Studies* (New York: McGraw-Hill, 1964), pp. 45–48.

pieces, spaced about a foot apart, whose ends are attached to sturdy upright poles which also support a sheltering roof over the drying crop.[15]

Once it has been cured, hay may be stored loose or in bales. Ranchers in some parts of the western United States have enormous hay-stacking machines, which can pile up a field stack containing several tons of loose hay in a very short time. Square, flat-topped stacks are found principally in California; high, round-topped stacks are common in the intermountain states; and low, round-topped stacks are made on the Great Plains.[16] Field stacks of hay are also common in parts of the Southeast, where they are built around a central pole for stability, but in the Middle West most farmers prefer to store loose hay in the loft, or hay-mow, of a barn. Hay bales, which are a bit more weather-resistant than loose hay, may also be stacked in the fields, or they may be stored in any convenient building; on some farms in the South, where a changeover has been made from cotton to cattle, I have even seen baled hay stored in former sharecropper shacks.

Silage, for most Middle Western Americans, conjures up visions of chopped green corn stored in tall, vertical cylinders made of wood or concrete, but crops other than corn may also be ensiled. A horizontal silo consisting of a trench or pit dug into the ground, or an above-ground bunker constructed of concrete or treated lumber, may serve the same function as a vertical silo.[17] Corn is still by far the leading silage crop in the United States, accounting for almost three-quarters of all silage made in 1963, but sorghums have become increasingly important in drier areas, such as the southern Great Plains, and silage is also made from a variety of other products, including grasses, sugar beet tops, and the byproducts of fruit and vegetable processing plants.[18] Vertical silos still outnumber the horizontal ones, but the latter are increasing far more rapidly, because they are cheaper and easier to construct, and they are readily adaptable to self-feeding.

Where new cylindrical silos have been constructed, they are often in a feedlot, rather than beside the barn, and they are equipped with automatic unloaders which deliver silage directly into the feed troughs at the touch of a button. A recent innovation in silage technology has been the development of a proprietary glass-lined metal silo, which preserves silage much as glass jars are used by a housewife to preserve fruits and vegetables. Although these silos preserve fodder in almost the condition in which it was harvested, they are appreciably more expensive than conventional models. Apart from their initial expense, many farmers

[15] Richard Weiss, *Häuser und Landschaften der Schweiz* (Erlenbach-Zürich: Eugen Rentsch Verlag, 1959), pp. 252–56; this book is a powerful argument for the retention of the German language examination for the Ph.D., but even those who cannot read German can appreciate its splendid illustrations.

[16] W. H. Hosterman, *Measuring Hay in Stacks,* Leaflet No. 72 (Washington, D.C.: U.S. Department of Agriculture, 1931).

[17] J. R. McCalmont, *Farm Silos,* Miscellaneous Publication No. 810 (Washington, D.C.: U.S. Department of Agriculture, 1963).

[18] Paul E. Strickler, Helen V. Smith, and James R. Kendall, *Silos, Silage Handling Practices, and Minor Feed Products,* Statistical Bulletin No. 415 (Washington, D.C.: U.S. Department of Agriculture, 1968).

object to the big blue metal silos because of their visibility, which is quick to catch the eye of the tax assessor.

Many farmers have become increasingly reluctant to trust animals to harvest their forage crops, and prefer to do the job themselves, because they can do it more efficiently. The grazing animals waste some of the crop by trampling it, and soil a surprising amount of it with their droppings. They have to graze it over a period of time, consuming some when it is still green and immature, and some when it is overripe and past its prime. The farmer can harvest the entire crop at just the right time, store it in his silo, and dole out the proper amounts in his feedlot troughs. Many farmers, even in such traditional grazing areas as the Upper Lake States dairy country, have begun to switch to feedlot feeding year-round, and now harvest their crops with forage harvesters rather than with cattle. The animals never set foot on pasture, and farmers have begun to remove their permanent fences, which are no longer necessary; if he needs to enclose a piece of ground temporarily, the farmer can "slap a hot wire" (electric fence) around it.

GRAIN STORAGE. Part of the United States grain crop (10 to 20 percent of the corn, for example) may be carried directly from the harvest field to the local grain elevator and sold. If all farmers sell their grain at harvest time, however, the price is likely to be depressed, and some farmers have invested in facilities for storing their grain until the price is right. Furthermore, the bulk of the grain crop is produced to be fed to livestock during the winter months, not to be sold, and most farmers must provide storage both for the grain they produce and for any they have to buy to supplement their own production. Much of the grain is stored in bins in other buildings, especially barns and corncribs, but round corrugated-metal bins set on their own individual slabs of concrete have become a common sight in many rural areas as yields have increased and mechanical grain handling equipment has improved.

Better grain handling equipment has also produced changes in corncribs in the Middle West. The traditional crib had a lean-to roof, and open spaces were left between the boards on the sides so that air could pass through and dry the corn. Its height was restricted to about 12 feet in the days when ears of corn had to be husked by hand and loaded into the crib with a scoop shovel, and its capacity could be increased only by making it longer. Mechanical elevators enabled farmers to lift grain to a much greater height with less effort, and permitted them to construct taller cribs. Two cribs with lean-to roofs were built facing each other at a distance of ten to 20 feet, and the two roofs were extended to form a single gable roof. The upper portion of the central area was enclosed to form a granary, and the ground level could be used to store machinery and equipment. Both the granary and the cribs on either side could be filled through a cupola on the roof.

PROCESSING AND SALES. Fodder crops produced on the farm may be fed directly to livestock, and grain crops may be sold directly or fed, but some farm products require further processing before they are ready for sale or consumption. This processing commonly requires a distinctive structure, which may or may not be located on the producing farm. For some products this is no more than a level area where they can be spread out to dry, but many require a more elaborate structure.

Tobacco is a good example of a crop which must be processed on the farm before it can be sold. Tobacco leaves are still green when they are harvested, and they must be "cured" before they are ready for market. The curing process may involve heating, smoking, or simply drying, but a distinctive kind of structure is used, or even required, for each different process.[19] Hops are another crop which must be processed on the farm, and the structures associated with hop production and processing add a distinctive flavor to the rural landscape in the south-eastern corner of England. The hop vines are trained to grow up strings attached to wires some 12 to 15 feet above the ground, which are strung between stout poles in the hop yard.[20] After they have been picked, the green hops are dried over kilns in picturesque oast-houses, squat cylindrical structures with steep conical roofs. The roof of the oast-house is capped by a revolving cowl, or ventilator, which has a vane to keep it pointed into the wind, thus creating a draft for the kiln.[21]

In some of the more inaccessible parts of the United States it has been customary to process corn on the farm in order to reduce its bulk and convert it into a product of greater value per unit of weight before transporting it to market. This custom has been especially strong in Appalachia, where it has been founded on the technical competence which the early Ulster Scot settlers brought with them from Ireland. Although this practice is eminently reasonable from an economic point of view, it has run afoul of the legal authorities, who have seen fit to tax the resulting product at a rate which seems quite inequitable to the producers. Because of the considerable friction which has resulted, it is difficult, and sometimes downright dangerous, to attempt to describe the on-farm structures associated with the practice.[22]

The off-farm structures and plants in which farm crops are processed tend to be somewhat larger than those on farms. They require a capital investment beyond the means of most farmers, although some of the larger farm operations may have their own plants. The processing plant customarily is located in the local service center, and the processor also serves a marketing function, because he buys the crop which he processes. The cotton farmer sells his crop at the cotton gin, the dairy farmer delivers his milk to the creamery, the vineyard owner sells his grapes at the winery, the sugar producer sells his beets or cane at the refinery, and the fruit or vegetable man sells his crop at the packing plant or cannery. Each of these structures, the cotton gin, the creamery,

[19] Hart and Mather, "The Character of Tobacco Barns," footnote 3.

[20] L. Dudley Stamp, *The Land of Britain: Its Use and Misuse* (London: Longmans, Green, 1948), p. 126.

[21] Cook and Smith, *English Cottages and Farmhouses*, footnote 8, pp. 103–5.

[22] Loyal Durand, Jr., " 'Mountain Moonshining' in East Tennessee," *Geographical Review*, Vol. 46 (1956), 168–81; and Eliot Wigginton, ed., *The Foxfire Book* (Garden City, N.Y.: Doubleday, 1972), pp. 301–45.

the winery, the sugar refinery, the packing plant, and the cannery, is a distinctive structure characteristic of the area where a particular crop is produced, although all are as likely to be found in the small towns as in the countryside.

Distinctive sales structures which do not involve processing are also characteristic features of the central places of different agricultural regions. Nearly every whistle stop on the railroad lines traversing grain farming districts seems to have its battery of grain elevators. Many small towns in livestock and truck farming areas have local auction barns, where regular sales are held in season. Market centers in tobacco producing districts have enormous sales warehouses, roofs studded with rows of skylights to illuminate the cavernous sales floor where the crop is auctioned off in a fascinating process. Central stockyards, where farmers sell their fat animals to meat packing companies, tend to be in larger places than the other structures mentioned here, but all are as much a part of the distinctive landscape of an agricultural region as are its barns.

MACHINE SHEDS. Shelters for machinery and equipment are among the least distinctive and most standardized structures on contemporary farms; they are also among the dullest. I hope that I am not being too much of an antiquarian when I suggest that this may be because they are the newest. The modern machine shed is a very efficient and very functional metal structure without windows; one long side is either completely open, or consists mainly of large doors, which facilitates movement of machinery. Although machine sheds are not completely alike, they all look as though they had been bought from the same mail-order house, and the feeling persists that "when you've seen one, you've seen 'em all," whether you happen to be in Washington, Maine, California, Florida, or anywhere in between. The differences seem to be mainly a matter of size. Howard Gregor has compared the equipment yards of large farms in California to military depots, and I find the analogy depressingly appropriate.[23]

An Essay on Barns

Barns are the largest, most impressive, and probably the least understood structures on farms. One of the hottest arguments I ever witnessed took place between two geographers, both knowledgeable in the ways of the land, who were looking at the same structure and debating whether it should be called a barn or a shed. The one whose roots lay deep in the hills of the Appalachian South was appalled that anyone could be so dismally ignorant as to fail to understand that such a large and imposing structure was a barn. The one who was born and raised in one of the richest corn counties in Iowa had nothing but scorn for anyone who could be so stupid as to suggest that such a "dinky" little old building was anything but a shed. And their discussion of each other's

[23] Howard F. Gregor, "The Plantation in California," *Professional Geographer*, Vol. 14, No. 2 (March 1962), 2.

intelligence, genealogies, and general observational abilities was conducted, mind you, while they were both looking at the very same building.

The fact that two competent scholars could stand in the same field, look at the same building, and disagree on whether or not it should be called a barn, serves to illustrate the enormous variations in the meaning of the term as it is used in different areas and on different continents. For many years I have been interested in the geography of barns, and I have tried to focus on types, or models, and to avoid becoming bogged down in the details which make almost every barn unique. Much of my knowledge has been gained by looking at barns in various areas, and talking to farmers about them, and this essay represents my present state of information, surmise, and conjecture. I should be very much surprised if all of my notions are accepted, but I present them in the hope that they will encourage fellow scholars to start taking a longer and harder look at an important type of landscape feature whose geography has been relatively neglected.

BARNS IN WESTERN EUROPE. The word "barn" is derived from the Old English word *bereaern,* which meant "barley place" (from *bere* barley + *aern* place). In Western Europe the name "barn" can only be used for the farm building in which grain is threshed and stored, and it is unthinkable that the barn should be used to house livestock; each type of farm animal has its own structure with its own name. In England, for example, any countryman can tell you that the barn is for threshing, and horses belong in the *stable,* cows in the *byre* or *shippon,* and pigs in the *sty.* In France you thresh in the *grange,* but keep horses in the *écurie,* cows in the *étable,* and porkers in the *porcherie.* In Germany, as nearly as I can tell (although lexicographers, urban types all, are not very helpful when it comes to making such distinctions), you thresh in the *scheune,* keep horses in the *pferdestall,* and cows in the *kuhstall.* The names *shippon* and *byre* failed to make it across the Atlantic. Even though *stable* and *sty* successfully weathered the crossing, in American usage the meaning of the word *barn* has been expanded to include almost any big old building on a farm which is used to house crops and/or livestock.

The modular barn of Western Europe in bygone years was a simple rectangular structure, high as a two-story house, with large double doors opposite each other on the two long sides (Fig. 8–1). One door opened into the rickyard, and the other into the stockyard. (see Fig. 3–3). At harvest time the sheaves of ripened grain were built into neat, thatch-covered ricks in the rickyard, or stacked in the ends of the barn. The grain was threshed with hand flails on the barn floor between the two doors; the doors could be opened or shut to regulate the draft which carried away the dust and chaff. Right through the winter two men flailed away each day on the threshing floor until they had produced enough straw to feed and bed the cattle in the stockyard. The grain not needed for feed was stored in the ends of the barn, which might or might not have lofts. The stockyard was the farm fertilizer factory; although the farmer might hope

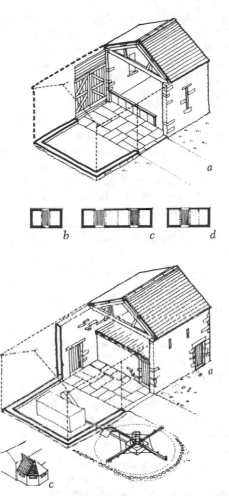

FIG. 8–1. *Cutaway isometric drawings of two English barns. The top draw-ing shows the standard barn of better farming areas; grain was threshed with hand flails on the central threshing floor, for which opposing doors provided light and a draft. Most barns had a symmetrical ground plan (cf. Fig. 3–3), but on larger farms barns might have two threshing floors or unequal end bays. French and English settlers brought this type of barn to New France, New England, and areas of English settlement along the Atlantic Seaboard; Yankees carried it westward into southern Ontario and southern Michigan. The one-level barn was modified in the New World by the addition of stalls for animals and haylofts in either end.*

    The bottom drawing shows one way in which such barns were modified when threshing machines replaced hand flails. Power for the threshing ma-chine might be provided by a windmill, a water wheel, a steam engine, or, as in this drawing, by horses plodding in circles; the power source was housed in a distinctive structure beside the bay of the barn which held the threshing machine. Both drawings are reprinted from R. W. Brunskill, Illustrated Hand-book of Vernacular Architecture *(London: Faber and Faber, 1971), 141 and 143, by permission of Faber and Faber, Ltd.*

to sell the fattened animals at a profit, and often did, their principal contribution to the farm was the manure they produced.

Although the basic module of rickyard, barn, and stockyard is quite simple, on many farmsteads it is obscured by a clutter of other structures, and it is especially well concealed if the farmstead is jammed in amongst the other buildings of a village, as so many are. The module can be seen most easily in the "field barns" which have been built on isolated parts of awkwardly shaped farms. Some of the farms resulting from enclosure, for example, were shaped like a slice of pie, with their apices at the old farm buildings in the village (see Fig. 3–3). The widest part of such a farm, at the "back end," might be as much as a mile and a half or more from the farmstead. In order to avoid having to haul sheaves of grain all the way to the farmstead, and then having to cart the manure back to the fields, the farmer built a field barn at the back end of the farm. Field barns consisting of a rickyard, a barn, and a stockyard, precisely as described here, are common features of the rural landscape in Western Europe, and they may be seen on many detailed topographic maps.

The majority of the barns in Europe are at least a century old, and many have been extensively modified, especially in the years since threshing machines replaced men with hand flails. On some farms the new threshing machine was permanently installed in one end of the old barn (Fig. 8–1). In the south of Scotland the threshing mill, as it is popularly known, may be driven by quite a variety of power sources, which are indicated by otherwise mysterious structures or projections on the back side of the barn. A long, narrow, projection parallelling the long side of the barn commonly houses an undershot water wheel, driven by a small stream whose diverted waters might be stored behind a dam in a tiny farm millpond. A small, blocky, projection might contain a coal-fired furnace for steam power, if it has a tall smoke stack, but it might also be the base of an old windmill. If the structure is round or octagonal, it probably sheltered a treadmill, and the threshing machine was operated, quite literally, by one horsepower.[24]

EUROPEAN BARNS AND FARMSTEADS.   At the risk of egregious over-generalization, the farmsteads of Western Europe might be divided into three broad categories as far as the placement of their barns is concerned. The farmsteads in the first group have no barns, because they are in areas so damp or so cool that oats is the only grain that can be produced, and little enough of oats. Both man and beast are sheltered under the same roof in a single small structure. The crop of oats, even if it ripens, is so meager that it can be threshed on the floor between the front and back doors at the living end of the house.[25] Examples, such as the "maison-bloc" of France, the "long house" of Wales, and the

    [24] J. A. Hellen, "Agricultural Innovation and Detectable Landscape Margins: The Case of the Wheelhouse in Northumberland," *Agricultural History Review,* Vol. 20 (1972), 140–54.
    [25] E. Estyn Evans, *Irish Folk Ways* (London: Routledge & Kegan Paul, 1957), p. 215.

traditional peasant cottages of the wetter parts of Scotland and Ireland, may be found only in the poorest and most primitive areas.

The farmsteads of the second group also consist of a single structure, with people, crops, and animals under the same roof, but the structure is vastly larger than the peasant cottage, and has far more internal differentiation. The living quarters, commonly several stories high and one room deep, occupy the front of the building. The central part is the threshing floor, and the back is divided into quarters for animals.[26] Today one finds this type of structure mainly in the northwestern marshlands and in the southern hills and mountains. In hilly areas many are built on slopes; the second level of the front is at ground level in the back, or may be reached from the ground quite easily by a gentle ramp.

The farmsteads of the third group have separate buildings, including separate barns, for separate farm functions. They may be divided into two major subgroups on the basis of their ground plans. In one subgroup the buildings are arranged around a central courtyard, commonly in some formal standardized order. Many farmsteads in Denmark, for example, have the farmhouse facing the barn across the courtyard, with the stable on the left and the byre on the right as you leave the back door of the farmhouse.[27] The farmsteads in the other subgroup also have separate buildings for separate functions, but these buildings are scattered around the farmstead with no apparent order or plan. Farmsteads of this type are common in England and in central Sweden.

NEW ENGLAND AND NEW FRANCE. On the basis of similarities in form and function, it is tempting to see European barn and farmstead types as ancestral to those of America. The first settlers in New England and New France built the same kinds of barns they had known in Europe, rectangular structures with a central threshing floor and room for grain storage at either end (see Fig. 8–1).[28] These were barns, not stables or

---

[26] The word "threshold" is derived from medieval English structures of this *Dreisassenhaus* type. A wide transverse passage separated the living quarters in the front from the animals in the back. This passage, which was used for threshing in season (hence "threshold"), was the main entryway to the building, with side doors leading front and back; Seebohm, *The Evolution of the English Farm*, footnote 9, pp. 94–95.

[27] Hart, "Vestergaard: A Farm in Denmark," footnote 14.

[28] Robert-Lionel Séguin, *Les granges du Québec du XVIIᵉ au XIXᵉ siècle*, Bulletin No. 192 (Ottawa: Musée National du Canada, 1963); and Fred Kniffen, "Folk Housing: Key to Diffusion," *Annals*, Association of American Geographers, Vol. 55 (1965), 558 and 563. The "connecting barn," one of those obvious things that geographers enjoy pointing out to each other as they tootle around the countryside, seems to have received more attention than it deserves; Wilbur Zelinsky, "The New England Connecting Barn," *Geographical Review*, Vol. 48 (1958), 540–53. Covered access to livestock has not been of any great importance to farmers on either side of the Atlantic, but "antecedents" for farmhouses connected to outbuildings are easy to find in Britain and on the continent; R. W. Brunskill, *Illustrated Handbook of Vernacular Architecture* (London: Faber & Faber, 1971); and Henry Glassie, *Pattern in the Material Folk Culture of the Eastern United States* (Philadelphia: University of Pennsylvania Press, 1968), pp. 185–87. In fact, Séguin says that early settlers in the St. Lawrence Valley brought the French *maison-cour*, or courtyard type of farmstead, with them from Normandy, but it failed to take root, and fell from favor in the nineteenth century (p. 31).

byres; the very idea of keeping horses or cattle in one of them would have made about as much sense to its owner as the idea of keeping horses or cattle in his living room or kitchen.[29]

The settlers began to modify their notions about their barns, however, as they adjusted their farming system to the climate. Hay became more important than grain, which did not ripen well in the cool damp summers, and both cattle and hay needed shelter during the winters, so both were brought into the barn, which was cheaper than building a completely new byre and a new hay shed. Stalls for the animals were placed in one end of the barn, and lofts for hay storage were built about seven feet above the floor in one or both ends.[30]

The simple, single-level, New England barn had cattle stalls in one end beneath the hayloft, a transverse threshing floor in the center, and grain or hay storage in the other end. This barn was carried westward to dairy areas in Michigan and Wisconsin, where it was still being built on small farms in the early twentieth century. The barn could be enlarged, if the dairy herd outgrew it, by extending the stall area at right angles to the barn, thereby changing the ground plan from an I-shape to an L-shape.

THE BASEMENT BARNS OF UPSTATE NEW YORK. The basement barns of upstate New York, which was settled by New Englanders in the late eighteenth and early nineteenth centuries, are quite unlike the barns of New England. The New England barn was basically a single level barn for threshing grain, with stalls for stock and a hayloft added as modifications. The New York basement barn had two distinct levels: the basement, commonly of stone, was for livestock, and the threshing floor and hayloft were on the upper level. In many barns the basement was partially excavated out of a hillside, so wagons could be driven straight onto the threshing floor from the uphill side of the barn; on level sites access to the upper level was provided by a gentle ramp, or barn bank.

I do not know where the idea for this basement barn originated. It is possible, but quite unlikely, that Yankees from New England suddenly got smart and invented it once they had crossed the Berkshires. The idea did not come from the Dutch.[31] Perhaps it was brought to this country by settlers from the English Lake District, where similar barns are common.[32] Perhaps it could be attributed to the small group of Palatine Germans who settled early in the eighteenth century in the lower

---

[29] Ennals has cited evidence which clearly demonstrates that Ontario farmers used their barns for threshing and left their cattle out over winter without shelter as late as the middle of the nineteenth century; Peter M. Ennals, "Nineteenth-Century Barns in Southern Ontario," *The Canadian Geographer*, Vol. 16 (1972), 256–70.

[30] Kniffen, "Folk Housing," footnote 28, p. 558; Séguin, "Les granges du Québec," footnote 28, p. 61.

[31] John Fitchen, *The New World Dutch Barn* (Syracuse, N.Y.: Syracuse University Press, 1968).

[32] Brunskill, *Illustrated Handbook of Vernacular Architecture*, footnote 28, pp. 138–43.

Mohawk Valley, an area through which westward-moving New Englanders had to pass.[33] Whatever its origin, the upstate New York basement barn followed the New England barn westward, and became the standard barn in dairy areas of Wisconsin and Minnesota.

THE PENNSYLVANIA BARN. The multifunctional farm structures of southern Germany and adjacent areas (in which men, crops, and animals were all housed under a single roof) were much better adapted to the North American winter than the scattered buildings of the English farmstead, and the barns which early German settlers built in southeastern Pennsylvania had an important influence on the development of barn types in the United States.[34] The Pennsylvania barn (locally known as a *Schweizerscheuer*, or Swiss barn) was larger and better built than the New England barn, and from the very beginning it housed both livestock and crops, but never people, unlike its European prototype.[35]

The Pennsylvania barn was a two-level structure, with all of the livestock on the ground level (Fig. 8-2). The upper level had a threshing floor, bins for grain, lofts for hay, straw, and unthreshed sheaves of grain, and ample storage space for farm implements and machinery. The upper level was reached by a gentle ramp, or barn bank, on one side, which led to a central transverse threshing floor. On the opposite side this level projected some three to six feet out over the stockyard in a distinctive forebay. The forebay ran the entire length of the barn, and was an integral part of its structure. The forebay was built on cantilevered joists, and might be supported by pillars under the joists, or by an extension of the end walls, but on many barns it was supported only by the joists themselves. The forebay sheltered a line of Dutch doors which connected the stockyard with the animals' quarters on the ground level; the upper half of these doors could be opened even in bad weather to admit air and light, while the bottom half remained closed to keep out driving rain and snow.

The projection of the forebay also helped to keep the doors clear of straw, which was thrown into the stockyard for feed after grain had been threshed on the upper level. The stockyard, as in Europe, was a manure pen where fertilizer was manufactured. The open storage of manure is wasteful, and some farmers, especially the Amish, protected their manure from the elements by adding a "straw shed" almost as

[33] The importance of these German settlers went far beyond their mere numbers, and they certainly were familiar with structures quite similar to the basement barn from their European homeland; D. W. Meinig, "The Colonial Period, 1609–1775," in John H. Thompson, ed., *Geography of New York State* (Syracuse, N.Y.: Syracuse University Press, 1966), p. 131.

[34] Kniffen, footnote 28, p. 558.

[35] Alfred L. Shoemaker, ed., *The Pennsylvania Barn* (Kutztown, Pa.: Pennsylvania Folklife Society, 1959), p. 6. For a superb description of farmsteads and their buildings in southeastern Pennsylvania, see Long, *The Pennsylvania German Family Farm*, footnote 2, pp. 314–59.

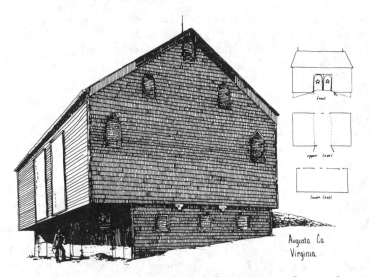

FIG. 8–2.   *A Pennsylvania Dutch barn. Livestock were housed on the lower level. A gentle ramp, or barn bank, led to the threshing floor on the upper level, which projected out over the barnyard in a forebay. Reproduced by permission of the author and publisher from Henry Glassie,* "The Barns of Appalachia," *Mountain Life and Work, Vol. 40, No. 2 (Summer 1965), 27.*

large as the barn itself, which changed the basic rectangular ground plan into a square or an L-shape.[36]

Some elements of the Pennsylvania barn have been copied more widely than others. The idea that crops and livestock both belong in the same barn, for example, is accepted as self-evident truth by most Americans, and barn banks may be found in many parts of the United States, but forebays and decorations remain distinctive features of the barns of southeastern Pennsylvania, and barn decoration is localized even within the area. Most of the barns with "hex signs" (so called for the benefit of people from New York City and other gullible tourists) are within a 20-mile radius of Kutztown, between Reading and Allentown; in the 1830s and 1840s, when local farmers began to paint their barns, they merely added these traditional designs to make them look prettier. Barns with brick-end decorations, highly stylized designs created by leaving selected bricks out of the brickwork in the end walls, are mainly south and southwest of Harrisburg. The resulting openings, which admit light and air, perform the same function as the narrow slits and loopholes which were made in the end walls of stone barns.

BARNS AND CRIBS IN APPALACHIA.   Although German settlers brought the knowledge of basic barn types (and of log construction methods as

[36] Walter M. Kollmorgen, *Culture of a Contemporary Rural Community: The Old Order Amish of Lancaster County, Pennsylvania,* Rural Life Studies No. 4 (Washington, D.C.: U.S. Department of Agriculture, Bureau of Agricultural Economics, 1942), p. 38.

well) to the United States, the dissemination of this knowledge depended on the restless, aggressive Scotch-Irish, frontiersmen par excellence, who made important modifications as they moved southward and westward into areas of very mixed quality. Those who settled on the better limestone lands of Maryland and Virginia continued to raise wheat and build Pennsylvania barns, even though the winters were milder and shelter for livestock was not as essential as it had been farther north. In the poorer hill areas, however, livestock shelters were dispensed with, wheat was replaced as the staple crop by corn, which did not have to be threshed, and the palatial barn gave way to a small corncrib of rough-hewn logs (Fig. 8–3).

Greene County Virginia

FIG. 8–3. *Appalachian corncrib. The simple rectangular log pen was the module from which a variety of barn types were developed in Appalachia (Figs 8–4 and 8–5). Reproduced by permission of the author and publisher from Henry Glassie, "The Barns of Appalachia,"* Mountain Life and Work, *Vol. 40, No. 2 (Summer 1965), 22.*

The simple rectangular log crib with one door in the gable end became the module from which a variety of barn types were developed in Appalachia by manipulating the number and position of individual cribs.[37] The simplest addition to the corn crib was a lean-to shed, in which tools and equipment could be stored. If a farmer prospered, he might build a larger crib, with a loft for storing hay, and lean-to sheds for livestock on either side (Fig. 8–4). The size of his crib was restricted, however, by the fact that a log structure cannot be much larger than 24 to 30 feet square, because the taper and weight of the logs limit the length which can be used. If the farmer needed more space he had to build two cribs, either side by side or facing each other; when he covered

[37] Kniffen, "Folk Housing," footnote 28, pp. 563–66; Glassie traces the basic constructional form back to the neolithic period in Germany; Henry Glassie, "The Old Barns of Appalachia," *Mountain Life and Work*, Vol. 40 (Summer 1965), 21.

Highland Co. Virginia

FIG. 8–4. *Appalachian single-crib barn with lean-to sheds, ancestor of the standard feeder barn of livestock areas in the Middle West (Fig. 8–6). Reproduced by permission of the author and publisher from Henry Glassie, "The Barns of Appalachia," Mountain Life and Work, Vol. 40, No. 2 (Summer 1965), 1.*

them with a common roof he created a double-crib barn. The runway between the cribs could be used to store machinery, or it might be planked over and used as a threshing floor (Fig. 8–5).

The ultimate development of the crib barn apparently originated in the rich limestone valleys of eastern Tennessee; it had four square cribs, one at each corner, with a complete loft above. Hay was stored in the loft, and the cribs housed corn and whatever animals were on the farm. At first the cribs were separated by runways, from side to side and from end to end, but in time the side runway was blocked off to provide more room inside the barn, and this created a structure with openings in the gable ends rather than in the sides. The farmer closed off the side runway instead of the end one because he did not need a threshing floor, and it was much easier to tack lean-to sheds onto the sides of the barn, rather than the ends, if he needed more storage space for animals and machinery.

Some students of barns have failed to grasp the importance of the association between wheat, threshing, and side entry. The shift from side openings to end openings signified a critical change in the basic function of the barn; it was no longer a structure for threshing wheat and other small grains, but a shelter for hay. Wheat had given way to corn as the principal grain crop, and the old hand flails had been replaced by new threshing machines. The threshing machine might be set up on the old threshing floor because custom dictated it, but the threshing floor itself was no longer necessary.

The end-opening barn, often with lean-to sheds on one or both sides, became the standard barn of the more prosperous farming areas of Appalachia. The vast hay loft, or mow, was filled by a hay fork mounted on a track which ran the full length of the ridgeline, and ex-

FIG. 8–5. *Transverse-crib barn with lean-to sheds. Reproduced by permission of the author and publisher from Henry Glassie, "The Barns of Appalachia,"* Mountain Life and Work, Vol. 40, No. 2 (Summer 1965), 29.

tended slightly beyond the end. The fork could be raised and lowered, opened and closed, and moved back and forth by a system of ropes and pulleys, which enabled the farmer to lift hay from a loaded wagon outside the barn and place it in the loft.

The gable ends are distinctive features of an end-opening hay barn. The projection of the hay-fork track beyond the barn was protected by an extension of the barn roof, which might be anything from a small peak to an enormous, boxlike structure whose sheer size appeared to threaten to tip over the entire barn. This structure sheltered the opening under the gables through which hay was lifted into the barn, but on many barns the function was performed more simply by a door which could be closed in bad weather.

BARNS IN THE MIDDLE WEST. When their ancestry, form, and function are considered, most barns in the Middle West belong to one of two basic types. The general purpose barn or feeder barn, which came from Appalachia, has end openings, lean-to sheds on one or both sides for livestock and/or machinery, and a large hayloft which may extend all the way to the ground in the center (Fig. 8–6). This is essentially a hay and corn barn which became the dominant type in livestock areas. One might have expected that settlers from the Middle Colonies would have brought their barn ideas with them to the Middle West, but the side-opening Pennsylvania barn is essentially a wheat barn, and in the Corn Belt it was eclipsed by the end-opening hay and corn barn when corn replaced wheat as the dominant crop.

The basement barn, or dairy barn, was introduced into the Middle

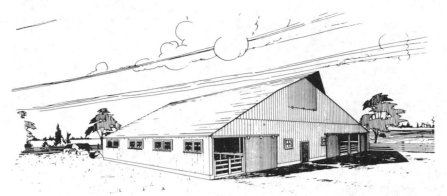

FIG. 8–6.   *Perspective drawing of a pole beef cattle barn, Plan 72432, Midwest Plan Service, Iowa State University, Ames, Iowa. In his letter granting me permission to reproduce this drawing, John H. Pedersen, manager of Midwest Plan Service, wrote "I hope the illustration will not be billed as one for modern beef production. Although there are many buildings around essentially like the plan, it hardly fits modern production systems."* Sic transit . . .

West from upstate New York.[38] Livestock are housed on the ground floor, which has masonry walls broken by numerous windows, and the upper level is a large hayloft (Fig. 8–7). Early barns of this type had side openings, with a barn bank leading to a central threshing floor on the upper level, but the combine harvester eliminated the need for a threshing floor, and many of the newer dairy barns have end openings. On some of the older barns the barn bank, which had become a relict feature, has been removed and replaced by a row of windows to admit more air and light to the ground level.

These two basic types have an enormous number of variants, because each farmer has his own ideas as to the kind of barn he needs, and sometimes these ideas have produced barns which do not fit neatly into either category. A quite distinctive and highly localized type of barn was built in the Kentucky Pennyroyal area in the 1920s, after the bottom dropped out of the dark tobacco market, and the former tobacco farmers who switched to dairy farming had to have barns for their cattle. The county agent in Barren County was an overwhelming personality, and he persuaded many farmers to build barns whose ground floors had sides which rose vertically to a height of three to five feet, then slanted outward and upward at a 30 degree angle for a few feet, and returned to the vertical some five to eight feet above the ground. The slanted side formed the outer wall of a manger into which hay could easily be pitched from the loft above. Barren County has many of them, but such slant-sided barns are few and far between outside this one county where the per-

[38] Loyal Durand, Jr., "Dairy Barns of Southeastern Wisconsin," *Economic Geography*, Vol. 19 (1943), 37–44.

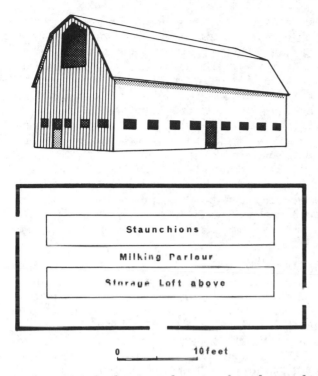

FIG. 8–7. *The basement barn, or dairy barn, which originated in upstate New York, and was carried westward by dairy farmers across southern Ontario, Michigan, and Wisconsin into Minnesota. Livestock are housed on the ground floor, which normally has solid masonry walls, and the upper level is a large hayloft. Most dairy barns have a towering cylindrical silo or two at one end. Reproduced by permission of the author and editor from Peter M. Ennals, "Nineteenth-Century Barns in Southern Ontario," The Canadian Geographer, Vol. 16 (1972), 263.*

suasive county agent held sway, and most of those few have been built by former residents of Barren County.

The round or polygonal barn is the most obvious of the unusual barn types.[39] The first big round stone barn in the United States was built at Hancock, Massachusetts, in 1825, but round barns had their greatest popularity during the great barn-building spree of the late nineteenth and early twentieth centuries. They were championed by some of the "improving" writers in the agricultural press because they combine maximum floor space with minimum wall space, but the construction of a round barn required special skills. The basic round barn has a silo in the center, livestock on the ground level, and a hayloft on the upper level, which is reached by a barn bank. Most farmers who have round barns

[39] *Time,* 4 July 1969, pp. 54–55.

do not seem to like them very much; they give a variety of reasons for their dislike, but most of these reasons can be boiled down to the simple fact that our rectangular society has no place for a round structure.

The shape of its roof is one of the most immediately obvious features of any barn, but thus far I have ignored differences in barn roof types because these differences are more closely related to the date of construction than to the type of barn.[40] The straight gable roof is the oldest form, the simplest to construct, and the most awkward to live with, because the low headroom under the eaves requires a good deal of uncomfortable bending and stretching in the hot loft (Fig. 8–6). Around the end of World War I the gable roof was largely replaced by the gambrel type, which provided much greater headroom (Fig. 8–7). The gambrel, in turn, was replaced by the fornicate, or Gothic, roof around 1940. It is not at all uncommon, of course, to see a new addition with a gambrel or fornicate roof attached onto the end of an older barn with a gable roof.

THE LAST OF THE BARN. Most of this essay on barns has been written in the past tense, because barn types began to be standardized and homogenized by the institutionalized designs which started pouring out of agricultural experiment stations a generation or so ago, and now the barn has been completely replaced as a functional form by a variety of new structures which are better designed to serve specific purposes on the individual farm. On a modern farm the barn is a relict feature; the hayloft is obsolescent, the threshing floor is obsolete, and who needs stalls for horses? Of course some use can be found for the old building, if it is still structurally sound, but as soon as it start to deteriorate the best thing to do is to pour kerosene on it and light a match. An essay on barns must close on a melancholy note, because one of the most impressive structures of the rural landscape is rapidly disappearing, and the examples which still remain deserve careful study while they are still with us.[41]

[40] Kiefer, "An Agricultural Settlement Complex," footnote 2, p. 496, Table 5.
[41] Although barns are fast disappearing from our countryside, books about barns have begun to appear on our coffee tables. Old barns, unfortunately, have seduced some architects, artists, and other aesthetes (who don't know apples from asparagus about agriculture) into preparing superbly illustrated, handsomely produced, and fearfully expensive compendia of misinformation.

CHAPTER 9

agricultural regions and farmsteads

Farmsteads are one of the major man-made ingredients of the rural landscape. Each farmstead reflects the activities of its farm, and each type of farming has a distinctive type of farmstead; a dairy farmstead, for example, is quite unlike a cash-grain farmstead, and either one would be out of place on a cotton farm. Farmsteads look pretty much alike over fairly large areas, because types of farming are similar over fairly large areas. This similarity of farmsteads and types of farming is one of the results of the agricultural regions created by the fact that large numbers of farmers have made similar decisions in response to similar conditions. Few agricultural regions are perfectly homogeneous, but most of them have enough similarity so that we can identify and describe the farmstead which is characteristic, typical, and most common in the region.

Delimitation of Agricultural Regions

Four approaches to the delimitation of agricultural regions have been more confusing than helpful:

1. at one one time it was fashionable to divide areas into "natural districts" and to describe farming within each district; British authors favored surface features as the basis for their divisions, whereas Americans leaned toward climate; [1]

[1] H. C. Darby, "Some Early Ideas on the Agricultural Regions of England," *Agricultural History Review*, Vol. 2 (1954), 30–47; and Derwent Whittlesey, "Major Agricultural Regions of the Earth," *Annals*, Association of American Geographers, Vol. 26 (1936), 199–240. Whittlesey properly lambasted earlier authors for "forcing agricultural regions into the alien pattern of climatic distribution" (p. 200), and then fell into the very same trap when he proclaimed that "from the geographic

2. some geographers have failed to distinguish between the way in which the land is used, which is clearly visible, and the integration of several different kinds of land use into an organization for economic production; the land may be used to grow trees, or grass, or crops of corn and soybeans, but these are no more types of farming than dairying, cash-grain farming, or beef production are forms of land use; [2]

3. some scholars have attempted to include entirely too many criteria in their classification systems; detail is useful and desirable, but a system which can produce the classification $4 \times V_3$ ($f_1$mz, lc + i_1 ol) + A_1 (cd + sm)lm is not likely to have widespread applicability; [3] and

4. some scholars have made their categories too fine; an attempt to distinguish "grain with roots" farms from "roots with grain" farms may be useful for some purposes, but it probably does not contribute appreciably to a better understanding of rural eastern England. [4]

An agricultural region, if it is to serve the purposes of description and generalization, should be a large (probably subcontinental) area, it

---

standpoint, Mediterranean agriculture is the most satisfactory of all types" (p. 226). No fewer than four quite different associations of crops and livestock were forced into this completely unsatisfactory category based on climate, yet it remained unchallenged in *Goode's World Atlas,* 13th ed. (Skokie, Ill.: Rand McNally, 1970), pp. 34–35.

[2] It is obviously a mistake to include "dairying" and "poultry" as categories on the map entitled "Crop Combinations" in John H. Thompson, ed., *Geography of New York State* (Syracuse, N.Y.: Syracuse University Press, 1966), p. 210, but failure to distinguish between land use and economic organization provided a more subtle source of confusion in J. W. Birch, "Rural Land Use: A Central Theme in Geography," in *Land Use and Resources: Studies in Applied Geography (A Memorial Volume to Sir Dudley Stamp),* Special Publication No. 1 (London: Institute of British Geographers, 1968), pp. 13–28. Presumably no one would question the desirability and importance of analyzing "the manner in which farms are functionally organized as a resource system with definable spatial and ecological relationships, both within and without the system" (p. 16), but the mind boggles at including such an analysis under the rubric of land use. Five basic categories of land use (cropland, grassland, woodland, built-up land, and other land) are generally recognized by serious students of the rural landscape on whichever side of the Atlantic; L. Dudley Stamp, *The Land of Britain: Its Use and Misuse* (London: Longmans, Green, 1948), pp. 22–24; *Basic Statistics of the National Inventory of Soil and Water Conservation Needs,* Statistical Bulletin 317 (Washington, D.C.: U.S. Department of Agriculture, 1962), pp. 9–10; and James R. Anderson, Ernest E. Hardy, and John T. Roach, *A Land-Use Classification System for Use With Remote Sensor Data,* Circular 671 (Washington, D.C.: U.S. Geological Survey, 1972).

[3] J. Kostrowicki, "Geographical Typology of Agriculture, Principles and Methods: An Invitation to Discussion," *Geographia Polonica,* Vol. 2 (1964), 158–67; and *idem,* "An Attempt to Determine the Geographical Types of Agriculture in East Central Europe on the Basis of the Case Studies on Land Utilization," *Geographia Polonica,* Vol. 5 (1965), 452–98; formula from p. 487. Kostrowicki's formidable checklist of variables should not be overlooked by any student of the rural scene, and each variable is well worthy of study in its own right, but their geographic variations are almost completely unrelated, and it probably will be difficult to integrate all of the variables in this Polish system into one single grand and glorious final typological synthesis which will serve both scientific and practical purposes.

[4] Each farm can become a separate agricultural region if a classification scheme is so detailed that it focuses on particular aspects of individual farms. Distinguishing "grain with roots" from "roots with grain" farms, and "cash roots and horticulture" from "cash cropping and horticulture" farms produced no less than 11 different farm types in an area about the size of the state of Massachusetts; B. Gyrth Jackson, C. S. Barnard, and F. G. Sturrock, *The Pattern of Farming in the Eastern Counties,* Occasional Papers No. 8 (Cambridge, England: Farm Economics Branch, Cambridge University, 1963).

should be defined by as few criteria as possible, and these criteria should be related to its inherent agricultural characteristics. The most useful agricultural regions probably are those which have been defined in terms of the main enterprise, or the principal source of income, of the individual farms within it. No single person could hope to be able to collect information from enough farmers to enable him to delimit agricultural regions over any large area. Energetic scholars have collected the data necessary for excellent studies of local areas by interviewing individual farmers, but at the subcontinental scale we are captives of the data which are collected and published by national census agencies.[5]

Data on the value of farm production are not collected in Britain, and British scholars have devised a variety of ingenious schemes for inferring the principal farm enterprise from other kinds of data. The pioneering effort was based primarily on land use, and divided English farms into cropland, grassland, and intermediate types, with five or six subdivisions of each type.[6] In Scotland the dominant enterprise of each farm was estimated on the basis of its total labor requirement; the number of acres under each crop, and the number of head of each kind of livestock, were multiplied by a standard man-hours-per-year value for that crop or animal, and the contribution of each to the total labor requirement of the farm was calculated.[7] In eastern England the value of gross farm output was substituted for total labor requirement, and a standard value for each crop and each kind of livestock was estimated by multiplying its average yield by its average price.[8]

Early attempts to identify agricultural regions in the United States were based primarily on the percentage of the land which was devoted to specific crops; Baker divided the eastern United States into a Spring Wheat Area, a Hay and Pasture Region, a Corn Belt, a Corn and Winter Wheat Belt, a Cotton Belt, and a Subtropical Coast.[9] Although Baker's map has been slightly modified through the years, it remains the standard, traditional map of the agricultural regions of the United States.[10] The indirect approach which Baker used has not been necessary since 1930, however, because in that year, and at each subsequent census, farmers have been asked to report the gross value of the various products which they have produced. In 1964 each commercial farm was placed in one of nine type-of-farm categories (cash grain, tobacco, cotton, other field crop, vegetable, fruit and nut, dairy, poultry, and other livestock) if its sales of the specified product or group of products amounted to 50 per-

[5] J. W. Birch, "Observations on the Delimitation of Farming-Type Regions, With Special Reference to the Isle of Man," *Transactions and Papers*, Institute of British Geographers, No. 20 (1954), pp. 141–58.

[6] Stamp, *The Land of Britain*, footnote 2, pp. 298–314.

[7] Department of Agriculture for Scotland, *Types of Farming in Scotland* (Edinburgh: H. M. Stationery Office, 1952).

[8] Jackson, *et al.*, *The Pattern of Farming*, footnote 4.

[9] O. E. Baker, "A Graphic Summary of American Agriculture, Based Largely on the Census of 1920," in *United States Department of Agriculture Yearbook, 1921* (Washington, D.C.: Government Printing Office, 1921), pp. 407–505; reference on p. 416.

[10] *Generalized Types of Farming in the United States*, Agriculture Information Bulletin No. 3 (Washington, D.C.: U.S. Department of Agriculture, 1950); and *Goode's World Atlas, op. cit.*, footnote 1, p. 70.

cent or more of its total value of farm products sold, but if no category produced half of its income the operation was classified as a general farm.

Virtually all statistical data-reporting units contain more than one type of farm, but the only information available for each unit is the aggregated data for all farms within it. The geographer who wishes to group these units into agricultural regions must devise some scheme which will enable him to classify each unit. In some units a given type of farm is clearly dominant, but in many the decision as to what constitutes dominance must be fairly arbitrary. Ideally, this decision should be based on the amount of land in the unit which is occupied by each type of farm, but in the United States this information is not available at the level of the county (the smallest unit of area for which agricultural data are published), and thus the grouping of counties into agricultural regions must be based on the number of farms of each type in each county.

The Middle West

The first step in defining an agricultural region is to define a farm. The official definition used by the U.S. Census Bureau is much too broad, and I decided that a census agricultural unit could not be considered a "real farm" unless it had sales of farm products which totalled $10,000 or more in 1964.[11] The counties which had at least one real farm per square mile in 1964 were concentrated, to a remarkable degree, in the Middle West; counties with at least half that number formed a ragged belt around them, with outlying pockets in the dairy areas of upstate New York and southeastern Pennsylvania, the Carolina tobacco country, the Delta and High Plains cotton areas, and the irrigated Central Valley of California (Fig. 9–1).

The concentration of agricultural counties in the Middle West was so striking that I restricted my attention to a block of 13 states in the nation's agricultural heartland (Fig. 9–2). I assigned each agricultural county to the type of farming category which was represented by the greatest number of farms in the county. A comparison with the traditional map of predominant types of farming in the United States reveals that my classification scheme greatly reduced the dairy farming areas in the north, almost eliminated the wheat and range livestock areas in the west, and completely wiped out the general farming area in the south, except for a few tobacco counties in Kentucky and the northern tip of the Delta cotton area in the Missouri bootheel.[12]

The Corn Belt was divided into a western section where livestock farms were most numerous, a cash-grain farming area in the center, and an eastern section where livestock and cash-grain farming counties

---

[11] John Fraser Hart, "A Map of the Agricultural Implosion," *Proceedings,* Association of American Geographers, Vol. 2 (1970), 68–71.

[12] "General farming" is a euphemism for "nonfarming" as far as most of the Appalachian and Ozark uplands are concerned.

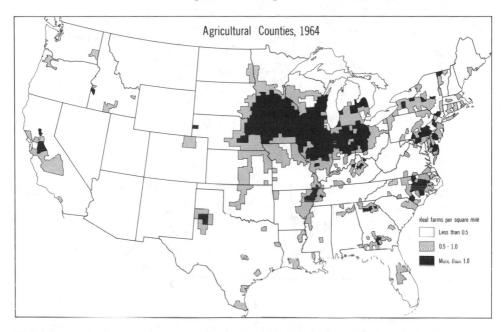

FIG. 9–1. Agricultural counties of the United States in 1964, based on the number of "real farms" per square mile. A "real farm" was defined as one which had a gross income of at least $10,000 from the sale of agricultural products. Reproduced by permission from the Proceedings, Association of American Geographers, *Vol. 2 (1970), 69.*

were intermingled. The surprisingly large number of cash-grain counties in the Corn Belt may be explained by the increasing specialization of individual farms. Many farmers have begun to concentrate on the single most profitable aspect of their traditional farm operation; some have dispensed with livestock and are concentrating on the production of cash crops, but others have begun to concentrate on feeding livestock, and they buy much of their feed from other farmers.

THE LIVESTOCK FARMING REGION. Iowa is the archetype of the Corn Belt (Fig. 9–3). On most farms the basic three-year rotation of corn, oats, and hay has been modified by practices such as continuous corn, growing two successive crops of corn, introducing soybeans as a cash crop, or leaving the hayfield a second year and using it for pasture. Corn, the principal crop, is primarily a meat-producing feed, and most of it is used to fatten hogs and beef cattle; sales of fat livestock provide the principal source of farm income. Oats and hay, which were never important "money" crops, have been giving way to soybeans as steadily increasing demand has kept pushing up the price of beans.

The kind of livestock which are fattened on any given farm will depend mainly on its size and on the amount of roughages it produces. As a general rule, cattle are kept on the larger farms, and on farms with

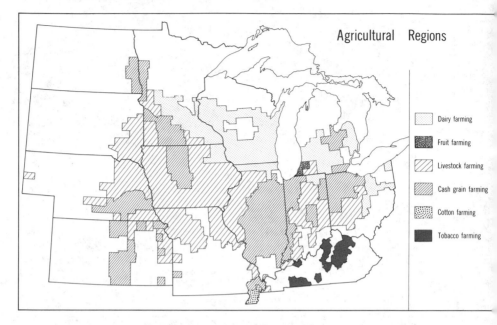

FIG. 9–2. *Agricultural regions of the Middle West in 1964. Each agricultural county (Fig. 9–1) was assigned to the type of farming category which was represented by the greatest number of farms in the county. Reproduced by permission from the* Proceedings, Association of American Geographers, Vol. 2 (1970), 70.

a good deal of land in pasture, whereas small farms and farms consisting mainly of cropland tend to specialize in hog production. Hogs usually have first call on the corn crop, because they are more efficient than cattle at converting concentrated feed into meat. A hog should be fat and ready for market in about six months, and it does not pay to buy lean hogs for fattening, so most hogs are raised on the same farm on which they are fattened. Hogs require more attention than cattle, and the man with a smaller acreage of cropland has more time to devote to his hogs.

Unlike hogs, most beef cattle are raised on one farm and fattened on another, because beef animals can effectively convert roughages (such as pasture and hay) into bone and hide, but they require concentrated feeds to flavor and tenderize their flesh. Most beef animals are raised on ranches in the dry range areas of the West, or on eastern farms with much rolling land whose steep slopes should be kept under permanent pasture or cultivated, if at all, only under a careful rotation system with considerable emphasis on grasses and legumes as cover crops. The lean animals, or "feeder cattle," produced on such farms must be fattened on corn before they are ready for market. A farmer whose summertime energies have been absorbed by his cropland will purchase feeder cattle in the fall, and he will fatten them during the winter when he cannot

LIVESTOCK FARMS AND RANCHES, 1964

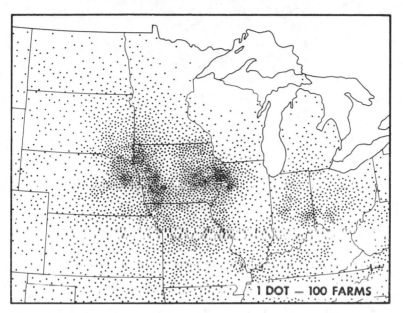

FIG. 9–3. Livestock farms in the Middle West in 1964. Reproduced from an original provided by the U. S. Bureau of the Census.

work in the fields. He is aware that a certain amount of the corn they eat passes through them undigested, and that hogs are not finicky about how and on what they dine, so he fattens two hogs "free" on the droppings left by each beef animal in his feedlot.[13]

Farms in the eastern portion of the livestock farming region, in Ohio and Indiana, were settled earlier and are smaller than farms in the western portion. Traditionally they have concentrated on production of corn and hogs, although cash-grain production has become increasingly important on the larger farm operations. The larger farms in the western portion, centered in Iowa, lie between the range livestock areas of the West, which are the principal source area for feeder cattle, and the major cities of the East, which are the main market for meat, so it is only logical that Iowa farmers should produce fat cattle as well as hogs and corn.

A typical corn-cattle-hog farming operation in Iowa involves so many different activities that the farmstead consists of a complex of buildings (Fig. 9–4). These buildings must provide shelter and storage space for corn, oats, hay, machinery, cattle, and hogs. Ear corn for livestock feed is stored in a double corncrib, with grain bins for shelled corn and oats above the central driveway; in recent years yields have increased

[13] Eugene Mather and John Fraser Hart, "The Geography of Manure," *Land Economics,* Vol. 32 (1956), 37.

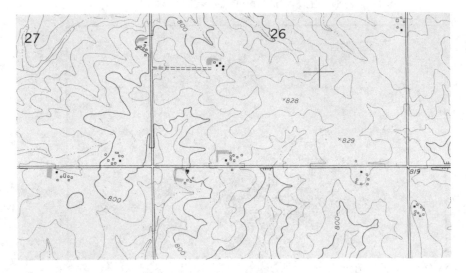

FIG. 9–4. A livestock farming area, as illustrated by part of the Tipton East, Iowa, 1:24,000 topographic quadrangle, which was published in 1953. The township had slightly more than four farmsteads per square mile, indicating a mean farm size of approximately 160 acres. Windbreaks have been planted on the north and west sides of some of these prairie farmsteads.

so much that a new circular metal grain bin has been added. Baled hay is stored in the lofts of the steer barn and the old horse barn, whose ground floor is now used for general purpose storage. A combination machine shed and workshop near the corncrib provides shelter and repair space for the equipment needed in crop production. Feeder steers are bought in October and fattened in the feedlot until around the middle of February; the steer barn opens onto the feedlot, and the animals can come and go as they please in any kind of weather. Some farmers have large permanent hog houses as part of the farmstead, but others prefer to reduce the risk of disease by having separate portable farrowing sheds, which are moved each year to the field which carries the hay crop.

The farmer is proud of the appearance of his farmstead, and the buildings are well maintained; the farmhouse is painted white, but the other buildings are "barn red." The farmhouse is completely air-conditioned, and has all of the comforts and conveniences which one would expect to find in the most modern suburban home. The orchard, the garden, and the hen house out back are all relics of an earlier day, when the farm wife raised much of her family's food; today she buys it at the supermarket, just as you and I do.

THE CASH-GRAIN FARMING REGION. Although farms in the cash-grain, or corn-soybean, farming region of central Illinois are roughly twice as large as those in the livestock farming region, the farmsteads are appreciably smaller and simpler, because soybeans have almost completely replaced small grains and hay in the traditional rotation, and cash-grain farmers concentrate on the production of corn and soybeans to the excusion of all else (Fig. 9–5). One man with modern machinery

CASH-GRAIN FARMS, 1964

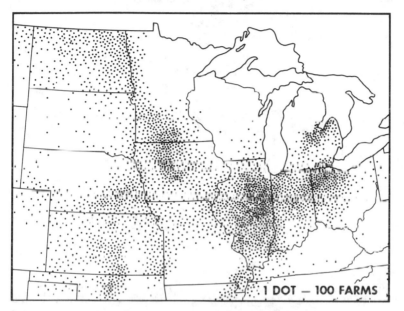

FIG. 9–5. *Cash-grain farms in the Middle West in 1964. Reproduced from an original provided by the U. S. Bureau of the Census.*

should be able to handle 200 to 300 acres each of corn and soybeans, but this keeps him pretty well occupied from April through October, and he does not feel that he can spare the time from crop production to operate an efficient livestock enterprise. Without livestock, however, he is underemployed during the winter months, and there is little to prevent him from enjoying a nice long vacation in the Florida sunshine. Envious neighbors, in fact, refer to cash-grain farmers as "B. C. M. farmers"; B for beans, C for corn, and M for Miami, where they spend the winter months.

A typical cash-grain farmstead on the Grand Prairie of eastern Illinois consists of a farmhouse, a double corncrib with grain bins for shelled corn and soybeans above the central driveway, and perhaps a combination garage-workshop-machine shed (Fig. 9–6). The house is attractive on the outside and livable on the inside, because the farmer and his wife desire and can afford all the creature comforts, but the lack of paint and disrepair of the weatherbeaten outbuildings indicate that neither the farmer, who is a tenant, nor his landlord is willing to spend very much money on them. The farmstead used to have a barn for

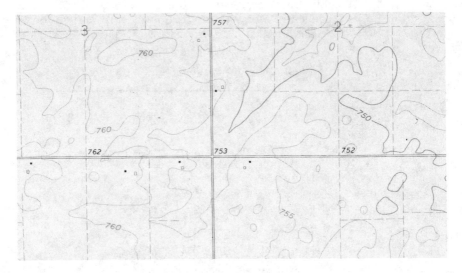

FIG. 9–6. *A cash-grain farming area, as illustrated by part of the Mount Gilboa, Indiana, 1:24,000 topographic quadrangle, which was published in 1962. The township had slightly more than three farmsteads per square mile, indicating a mean farm size of approximately 200 acres.*

horses, but about ten years ago the farmer poured kerosene on it and set a match to it, because the land on which it stood was too valuable to waste. The 24 horses it housed have been replaced by tractors, and the farmer boasts that every row on his farm is at least half a mile long, so that he will not have to waste too much time turning his large machines at the ends of the rows.

Various theories have been advanced to explain the existence of a cash-grain farming area in the very heart of the Corn Belt, where livestock might be expected to be more important. The land is remarkably level, which means that it is well suited to large-scale modern machinery, but is not liable to serious damage by erosion even if it is cultivated continuously under row crops. Many farms are operated by tenants, and perhaps neither landlord nor tenant is willing to invest in the buildings and fences which a livestock operation would require. Cities within the area have corn and soybean processing plants which provide a local market, and preferential freight rates encourage grain shipment by rail to other areas.

Arlin Fentem, after a careful analysis of the agricultural evolution of the Illinois Grand Prairie, concluded that the farms in this area produced a surplus of corn from the very beginning of farming, and he suggests that the favorable market and freight rate situation has been as much a response to this surplus as a cause of it. An important reason for the development of cash-grain farming in this area, he believes, was the expense of draining the remarkably level land, which required far more capital than most individual farmers could command, and thus greatly retarded settlement in the early days.[14] The job had to be done on a grand scale, and it was eventually accomplished by entrepreneurs who attempted to recoup their investment as quickly as possible by selling the newly drained land in large blocks.

From the very start, in other words, the farms of this area were large; so large, in fact, that most of the land had to be rented to tenants, and farm tenancy became accepted as a fact of life on the Grand Prairie. Cropping the land required the full energies of the farmer, who had no time for livestock, while good markets for his corn in the South, on the eastern seaboard, and in Europe provided him with an incentive for specialized cash-grain farming. Very low preferential freight rates were established in the 1860s and 70s by the railroads which were competing furiously for his business, and these rates have never been rescinded.

THE WHEAT FARMING REGIONS. West of the Corn Belt most farmers are prevented from practicing the traditional three-year rotation by scanty and unreliable precipitation, which restricts their production of corn and hay and leaves only the small grains, mainly wheat, as the dominant crop. Most of the cash-grain farmers of the Dakotas, Nebraska, and Kansas produce wheat rather than corn and soybeans. Wheat, in a sense, is the crop of default in dry lands; it produces better yields in areas which have more rainfall, but no other crop grows quite as well, or is quite as profitable, in semiarid areas. Furthermore, the semiarid portions of the Middle West are largely level to undulating plains which are very well suited to the large machines required for extensive wheat production; areas which are too rough for machine cultivation are used as range land for beef cattle.

Wheat farms are highly mechanized, and most of them are so large that relatively few wheat farming counties appear on the map of agricultural regions (see Fig. 9–2). About half of the year's work on a wheat farm is concentrated in the planting season, which lasts about a month, and another third is crammed into a 10- to 15-day harvest season.[15] For the rest of the year many wheat farmers seldom need to go near their land, and the bleak farmsteads of many wheat farms reflect the off-farm residence of sidewalk or suitcase farmers.

The occupied farmsteads are pretty much alike in the spring wheat area of the Dakotas and in the winter wheat area, which is centered on Kansas. A windbreak of scrawny conifers provides a bit of shelter against winds from the north and west, and only a creaking windmill breaks the farmstead's skyline. The farmhouse, an old horse barn with a small loft for hay, and the combination machine shed and workshop are small,

[14] The tardy settlement of the cash grain farming area in east central Illinois shows up quite clearly in John Fraser Hart, "The Middle West," *Annals*, Association of American Geographers, Vol. 62 (1972), map on p. 261.
[15] W. G. Heid, Jr., *Characteristics of Grain Farms with Emphasis on the Northern Plains Wheat Sector*, Bulletin 665 (Bozeman, Mont.: Montana Agricultural Experiment Station, 1973).

squat, and unpainted. The farmer relies, perhaps excessively, on the dry climate to keep rust away from his machinery, which often stands around out-of-doors, and only the battery of four glistening new metal grain bins gives any hint of prosperity.

THE DAIRY FARMING REGION. The Dairy Belt might best be thought of as a version of the Corn Belt which has been modified in response to the less favorable environmental conditions to the north (Fig. 9–7). Farm-

DAIRY FARMS, 1964

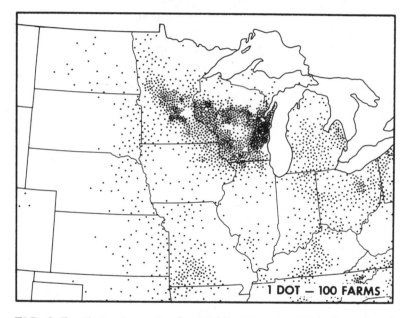

FIG. 9–7. *Dairy farms in the Middle West in 1964. Reproduced from an original provided by the U. S. Bureau of the Census.*

ers in the Dairy Belt follow the same traditional three-year rotation of corn, oats, and hay, but the forest soils of this region produce smaller yields of grain than the prairie soils farther south. The growing season is so short, cool, and humid that the ears of corn may fail to ripen properly, so the entire plant is cut and chopped while it is still green, and it is stored in silos for winter feed. Recent glaciation has left a rougher land surface, with more steep slopes, which both restricts the use of large machines and requires that more of the land be used for hay and pasture (Fig. 9–8). The farms of the Dairy Belt produce less concentrated feed, such as corn, which could be used to fatten livestock, but more roughages, such as silage and hay, which provide excellent sustenance for dairy cattle.

The most prominent structure of a typical dairy farmstead in Wis-

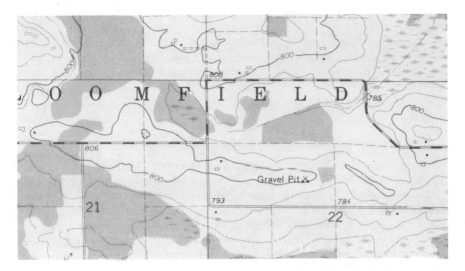

FIG. 9–8. A dairy farming area, as illustrated by part of the Poy Sippi, Wisconsin, 1:24,000 topographic quadrangle, which was published in 1961. The township had five or six farmsteads per square mile of cleared and drained land, indicating a mean farm size of approximately 120 acres. Steep slopes, gravel pits, marshes, swamps, and uncleared woodland are not uncommon in areas of recent glaciation.

consin is the distinctive dairy barn, with a large loft for hay storage, and a towering cylindrical silo beside it (see Fig. 9–8). The ground floor, where the cattle are kept tied to stanchions for much of the year, is built of masonry, and lined with windows to admit light and air. Next to the barn is a small milkhouse, where the milk is cooled and stored until the tank truck comes to collect it. Unless the farm has some enterprise supplementary to dairy farming, the only other structures in the farmstead are the farmhouse and a garage-workshop-machine shed, with bins above in which grain can be stored.

Dairy farmers in the Middle West produce most of their own feed, unlike dairy farmers farther east, where less favorable growing conditions are counterbalanced by proximity to major urban markets and a better price for milk, which enables the dairy farmer in the East to purchase feed concentrates instead of growing his own. As a general rule, the disposition of the milk, and the price the farmer receives for it, is tied fairly closely to the location of a farm in relation to an urban market. Milk from the closest farms supplies the home market for fluid milk, and receives the best price; milk from the farms a bit farther away goes to condenseries or cheese factories, and milk from the farthest farms goes to creameries to be made into butter. Many dairy farmers, especially those in the distant creamery-butter districts, supplement their milk checks by keeping a herd of hogs, a flock of laying hens, or a flock of turkeys. Others, in specialized districts within the Dairy Belt, grow cash crops such as dry beans, sugar beets, potatoes, and a variety of vegetables.

The eastern shore of Lake Michigan is a highly specialized fruit farming subregion within the Dairy Belt (see Fig. 9–2). The lake moderates the temperature of air masses which pass over it from the west, thereby retarding bud development in the spring until frost danger is past, and delaying the danger of frost until late in the fall.

Agricultural Regions and Rural Population

Although the distribution and demographic character of the rural population are, at best, only peripherally related to the look of the land, the interaction between the farming system practiced in an agricultural region and the nature of the farm population within that region is so significant that it warrants brief consideration. For example, it can provide the basis for a theoretical model which integrates agricultural systems, farm population densities, rural underemployment and over-population, and migration.

Given a certain level of technology, and enough time, one might expect some kind of equilibrium to develop between the farming system and the size of the farm population within each agricultural region, because the farm people provide the labor force for the farming system. No farmer can conduct a farm operation for which he does not have an adequate labor force, but conversely, if the farm population grows to exceed the size needed by farm operations in an area, then part of the farm labor force becomes surplus and is underemployed.

The number of farm workers who are required, and can be used effectively, at any given level of technology will vary greatly from one type of farm operation to another; truck crops, for example, demand more man-hours per acre than wheat, and dairy cattle require more labor than beef cattle. In any given agricultural region, however, a single type of farm operation tends to predominate, and the number of man-hours of labor required per acre of farmland fluctuates only within comparatively narrow limits.

The more-or-less standard number of man-hours required per acre of farmland in an agricultural region provides a basis for estimating the theoretical density of farm population in the region, if one may assume that variations in the density of farm population are a function of variations in farm size and variations in the size of farm families.[16] The number of man-hours required for each acre will largely determine the total number of acres of farmland which one man can handle, or the size of a one-man farm operation, and this size is more-or-less standard in a given agricultural region. The size of a farm family tends to vary only within relatively narrow limits in a given culture at a given time, and thus one might assume that each agricultural region will have a characteristic

---

[16] John Fraser Hart, "Some Comparisons of Population Distribution in the Middle West in 1950," *Proceedings,* Indiana Academy of Science, Vol. 71 (1961), 210–18.

density of farm population, which is nicely adjusted to the requirements of its farm operations.

For example, suppose that the "normal" farm size in a given region
was 160 acres, and the "normal" farm family consisted of four people;
four people per farm on four farms per square mile would produce a
"normal" farm population density of 16 persons per square mile. Now
hold one variable constant, change the other, and see what happens to
farm population density. If family size remained unchanged, but the
size of farm doubled, to 320 acres, the farm population density would
drop to eight persons per square mile; conversely, if farm size were
halved, to 80 acres, the farm population density would leap to 32 persons per square mile. If the size of farm remained 160 acres, but the normal family size went from four to five persons, the farm population density would go from 16 to 20 persons per square mile.

Despite the great theoretical attractiveness of the idea that each
agricultural region has a characteristic density of farm population, which
is in equilibrium with its farming system, this notion is based upon the
assumption of a static level of technology, which is not, and probably
never has been, valid in the United States. Constant technological change,
in fact, has been virtually a way of life on the American farm, especially
in the last half century or so. To cite only the most important example,
the introduction of the internal combustion engine put the farmer on a
tractor, multiplied the land area which he could work, and thereby greatly
increased the size of a one-man farm operation. This reduced the labor of
his sons to the status of a surplus agricultural commodity, and released
them from the land to swell the flood of migration from rural to urban
areas.

Migration can provide an agricultural region with a convenient safety
valve to relieve rural underemployment (if farms grow larger) or over-
population (if the labor force grows to exceed the requirements of the
farming system). Conversely, migration can also produce changes in an
agricultural region if it reduces the farm population below the level
necessary to support the existing farming system. A good example is the
large-scale migration of Negroes from the cotton plantations of the South
to urban areas in the North, which began shortly after the First World
War disrupted the flow of unskilled workers from southern and eastern
Europe. Cotton planters in the South, who had depended upon cheap
black labor, were forced to mechanize their operations or to shift to a
different form of production.[17]

Paradoxically, the labor requirements of some farming systems may
be strong enough to hold the rural population on the land. The enormous
amount of labor needed to produce flue-cured tobacco seems to be a
principal explanation for the fact that the tobacco counties of eastern
North Carolina have the densest rural population in the United States.[18]

---

[17] John Fraser Hart, *The Southeastern United States* (Princeton, N.J.: Van
Nostrand Reinhold, 1967), pp. 22–25.
[18] John Fraser Hart and Eugene Cotton Mather, "The Character of Tobacco
Barns and Their Role in the Tobacco Economy of the United States," *Annals,* Association of American Geographers, Vol. 51 (1961), 291–92.

These counties have been less affected by Negro out-migration than any other part of the rural South.[19] On the other hand, prolonged migration from a rural area (and especially the migration of young females, who can find little chance for employment in the agricultural economy) may eventually result in a rural population which is so old that it lacks the ability to reproduce itself. During the 1960s, for example, the number of deaths exceeded the number of births in a large block of counties in Missouri, Arkansas, Kansas, Oklahoma, and eastern Texas, and these counties experienced a natural decrease in population.[20]

In addition to the age structure and the rate of natural increase, other demographic characteristics of the rural population may be related fairly closely to the agricultural system of a region. In areas of large farms and sparse population, for example, the tax base is often too small to provide adequate support for such institutions as schools, libraries, and hospitals, and this inadequacy may be reflected in lower levels of educational attainment or health in the rural population.

Other areas, especially those which produce specialty crops under intensive management systems, may depend upon a cheap and servile labor force which is drawn primarily from a distinctive ethnic group. For example, cotton plantations in the South traditionally depended upon Negroes, either slaves or sharecroppers who were little better than slaves, to work the land, and as a consequence the population of many rural areas in the South has contained a relatively sizable Negro component. The importance of seasonal migrant workers in many intensive truck and fruit farming areas may not be evident from the columns of the census, which is taken at a time of the year when they are working someplace else, but even in their absence they impart a distinctive touch to the rural landscape.

Finally, each agricultural region requires a network of collecting, processing, and distributing centers to provide services and goods for its people. The spacing of towns and villages within the central place network is inversely related to the density of the rural population it serves.[21]

[19] John Fraser Hart, "The Changing Distribution of the American Negro," *Annals,* Association of American Geographers, Vol. 50 (1960), 263–64.

[20] Calvin L. Beale, "Natural Decrease of Population: The Current and Prospective Status of an Emergent American Phenomenon," *Demography,* Vol. 6 (1969), 91–99.

[21] Gerard Rushton, "Map Transformation of Point Patterns: Central Place Patterns in Areas of Variable Population Density," *Papers of the Regional Science Association,* Vol. 28 (1972), 111–29.

CHAPTER 10 *house types*

and villages

Houses are the most important structures (certainly for the people who dwell in them, if for no one else) in any rural area. Both the form of houses, and their arrangement on the land, are of interest to the student of the rural landscape. The study of house types is a central theme in cultural geography, because no other structure provides better insight into the folk culture of an area, yet no other theme is more complex, because the form of a house may be so completely divorced from its function. The basic function of a house, to protect its occupants against the rigors of climate, may be served by many different forms; in selecting any one particular form, a man is able to demonstrate his artistry, advertise his beliefs, and flaunt his wealth. Perhaps nothing tells so much about the values a man holds as the kind of house in which he chooses to live.

The description and explanation of the distribution of houses on the land is the core (and, according to some enthusiasts, the entire content) of settlement geography. The individual houses may be isolated one from another, or they may be clustered into hamlets, villages, towns, cities, and metropolises. Interest may focus on the layout of houses and other structures within clusters, or on the functions of clusters as central places providing goods and services for tributary areas, and on the relationships between clusters. Urban geographers, especially in recent years, have tended to concentrate their attention on the nonresidential aspects of central places, but within any given center, no matter how large or how small it may be, the most numerous functional structures are the houses in which people live, and they occupy the largest proportion of its total area.

House Types

HOUSES AND NATURE.   Other scholars have examined house types in
such detail that it is inappropriate for an amateur to do more than make a
few general observations. The first requirement of any dwelling is to provide shelter against the more serious environmental stresses. No living
thing, whether plant or animal, tries to settle in an area where it is too uncomfortable. Plants are the captives of their environments, but animals,
and especially the human animal, are able to exist with at least moderate
comfort in an astonishing range of environments by fashioning some kind
of shelter for themselves. Most primates have homes of some sort, and it
seems quite reasonable to assume that man, as the only naked primate
(and thus most susceptible to bad weather) and one whose offspring are
completely immobile and helpless, has had some kind of nest or den from
his earliest days.

"There can be hardly any doubt that man sought out the shelter of
a natural roof wherever he could. Under an overhang of rock he might
be out of the wet and wind, with a dry floor to sleep and sit on, and with
less nightly chilling of the air than in the open. Such shelters provided
the preferred, perhaps the first, home in any clime and became virtually
a necessity when fire was kept." [1] The roof, which freed man from the
rock shelter, may have been one of the first human inventions; it certainly
forced him into an awareness of the potential of his environment, because
even the simplest roof requires knowledge of the use of several different
kinds of plant material: poles to support it, leaves or grass to cover it,
and vines to bind it together.

Before cheap modern transport became available, all men were
heavily dependent upon the resources of the local environment for their
building materials, whether they used snow, stone, earthen matter (bricks
or adobe), parts of plants (sod, grass, or wood), or even the pelts of
animals. In the forested areas of western Europe men built their homes
of wood until the forests were decimated, and then they turned to stone
or brick. Stone of appropriate workability and durability, which had been
used since antiquity in areas where it was convenient, became the building material of prestige, but brick was used by the common folk.[2]

A consideration of the relationship between building materials and
the local environment, however, must be wary of entrapment by the
doctrine of retrospective inevitability. "With the settlement of North
America," asked Emrys Jones, "what was more natural than that use
should be made of the timber which had to be cleared even before crops

---

[1] Carl O. Sauer, "Sedentary and Mobile Bent in Early Societies," in Sherwood
L. Washburn, ed., *Social Life of Early Man*, Viking Fund Publications in Anthro-
pology No. 31 (New York, Wenner-Gren Foundation, 1961), p. 264.
[2] An excellent description of the building materials which have been used in
English villages is in Victor Bonham-Carter, *The English Village* (Harmondsworth,
England: Penguin, 1952), pp. 102–21. A superb treatment of building materials in
Switzerland is in Richard Weiss, *Häuser und Landschaften der Schweiz* (Elenbach-
Zürich: Eugen Rentsch Verlag, 1959), pp. 35–100.

were grown?" [3] What actually was more "natural," of course, was that the Englishmen who came to North America would keep right on building the same kinds of half-timbered wattle-and-daub houses they had been building in England, as is illustrated by the superb reconstruction of the first English settlement at Jamestown, Virginia; the construction of log cabins had to await the arrival of German settlers, who were accustomed to building entirely in wood. Incidentally, the question asked by Jones, if transplanted to the stony glacial soils of New England, conjures up rather startling visions of Cape Cod cottages built of field stone.

THE EVOLUTION OF HOUSE TYPES. A type of house, like any other culture trait or complex, originates at a certain time in a particular locality in response to local needs, local materials, and local technical skills. It gains acceptance, becomes elaborated, and is communicated outward, or diffused, to other areas by migration and emulation until it encounters sufficient resistance from unsuitable physical conditions, from superior types, or from disparate cultural levels. Most houses outlive the men who built them, and in a stable peasant society the house comes to be associated with ancestors. Its size and style tend to become sacrosanct, and its archaic symbols and meaningless details are repeated through the centuries. Not all folk tradition, of course, is illogical. The superstitious Irish, for example, believe that it is unlucky to build a house of more than a certain width; in the older days indeed it was, because a wider roof would have collapsed.

Throughout human history until the last century or so, and in much of the world even today, most areas have had their own distinctive regional house types, whose size, shape, and style are deeply imbedded in folk tradition. One of the curious traditions of western European culture, for example, holds that houses must be rectangular rather than round, although neither shape is dictated by the available building materials. Round houses may still be found in various parts of Europe, but they are oddities, and contemporary round construction is reserved almost entirely for ceremonial structures, such as the domes of cathedrals, capitols, and county court houses. Anyone can immediately think of a host of reasons why a round house is impractical, but on further thought it becomes obvious that these reasons are really products of the rectangular tradition, and not its cause.

As a general rule, the house types of economically more advanced societies are less closely related to their builders, their occupants, and their environments, than are those of simpler folk. With a stable society, and given enough time, most regions evolve traditional, vernacular, "anonymous," house types which blend the requirements and the resources of the environment with the culture of the people, but increasing mobility and a wider exchange of ideas seem to lead to the replacement of these regional folk-house types by standardized national types. More than half of the housing units in the United States have been constructed since the

[3] Emrys Jones, *Human Geography: An Introduction to Man and His World* (London: Chatto and Windus, 1964), pp. 112–13.

end of World War II, mainly in monotonous suburban developments of homogenized house types whose plans reflect the dominant national fashion. (National fashion can dictate a style which is locally incongruous; just ask any Minnesotan who has spent a winter shoveling snow to prevent the collapse of the flat roof of his ranch-style house.)

Sears, Roebuck, and Company influenced house types over a large part of the United States in the early part of this century.[4] In 1908 the Sears Spring Catalog introduced a Modern Homes Department, which sold complete homes and made mortgages on them anywhere east of the Mississippi and north of the Ohio. Sears had to reacquire a number of these properties during the Depression years, and in 1934 the Modern Homes Department was liquidated, but for a quarter of a century anyone could buy a complete new house of a standard style from Sears. ("Dear Sir: For enclosed check, please send me one house.") Several corporations bought enough houses for entire new subdivisions at Carlinville and Wood River in Illinois, at Chester and Plymouth Meeting in Pennsylvania, and at Akron, Ohio.

THE CLASSIFICATION OF HOUSE TYPES. Classification of house types in the field is only slightly less difficult than moving a cemetery or changing a curriculum, and brave indeed is the scholar who attempts it. "Neither houses nor other cultural forms can be classified in a manner exactly analogous to that used by biologists," said Fred Kniffen. "The biologist never finds the tail of a lion grafted to the body of a cow; the classifier of cultural forms has no such assurance."[5] Within certain limits, every man who builds a house is free to borrow and combine any elements which happen to appeal to him. In a traditional society these limits may be quite narrow, and it is in such societies that house types and their classification appear to be simplest; in the contemporary United States, however, every conceivable permutation (plus some which can only be described as inconceivable) seems to exist.

The most effective approach to the classification of house types would be to recognize and define the basic types which exist in an area, and then to examine the way in which these basic types have been modified and combined in individual houses. Unfortunately, however, the basic house types in the United States (if indeed such types actually do, or ever did, exist) have never been identified and defined in unambiguous terms, and in each area the scholar must determine for himself what the basic types are. His only data are the houses he sees before him, each of which appears to be unique unto itself, and the apparent chaos he encounters may be more real than apparent.

Under such circumstances, it is expedient to identify and classify the individual elements which comprise the house (such as the roof, chimney, height, walls, windows, doors, porches, appendages, and floor plan) and to record their variations for a large number of houses in the

[4] I am grateful to Miss Adele Kozan, of the Archives Department of Sears and Roebuck, for giving me this information about Sears and Roebuck houses.

[5] Fred B. Kniffen, "Louisiana House Types," *Annals,* Association of American Geographers, Vol. 26 (1936), 180–81.

hope that some patterns of association will emerge.[6] In most areas, alas, such hope is forlorn, and an objective, quantitative approach to the classification of house types, by its very detail, tends to obscure the basic forms which it is intended to identify. After experimenting with such an approach, most students of house types have turned to classification schemes based on arbitrary, highly subjective, even mystical identification of the basic house types in the areas they have studied. A subjective classification scheme is not objectionable if it is based on a high degree of familiarity with the houses of the study area, but the scholar who develops one has the responsibility of analyzing his subjective impressions as carefully as possible so that he can reduce his types to their basic elements, and describe them in such a way that his scheme can be understood and used by other scholars.

The Morphology of Settlement

DISPERSAL VS. NUCLEATION. In addition to his interest in the form of houses, the student of the rural landscape needs to understand the way in which they are distributed across the land. Farm houses, for example, may be dispersed through the open countryside, in isolation one from another, or they may be nucleated into hamlets and villages. But what constitutes isolation, or the open countryside? Even the half-humorous definition saying that you are in the open country if you don't have to pull the shades before undressing is one which fails to take into account some rather startling variations in the latent exhibitionism of human beings, but no better definition will be attempted here. It is generally agreed, however, that most American farmers live on their own land in dispersed or isolated farmsteads, although agricultural villages have been established in New England, in Mormon areas, and in the Spanish-American Southwest. In contrast, agricultural villages are the rule in many other parts of the world, and dispersed farmsteads are exceptional.

Geographers have invested a good deal of time and energy in attempts to measure and describe patterns of farmstead dispersal and

---

[6] The best available discussion of the classification of house types is R. W. Brunskill, *Illustrated Handbook of Vernacular Architecture* (London: Faber & Faber, 1971). Another excellent discussion is Henry Glassie, "The Types of the Southern Mountain Cabin," Appendix C of Jan H. Brunsvand, *The Study of American Folklore* (New York: W. W. Norton, 1968), pp. 338–70, and much useful information is in Henry Glassie, *Pattern in the Material Folk Culture of the Eastern United States* (Philadelphia: University of Pennsylvania Press, 1968). Most of the better known works on house types in North America are listed in the voluminous footnotes of John E. Rickert, "House Facades of the Northeastern United States: A Tool of Geographic Analysis," *Annals,* Association of American Geographers, Vol. 57 (1967), 211–38, but for some inexplicable reason Rickert omitted Fred Kniffen, "Folk Housing: Key to Diffusion," *Annals,* Association of American Geographers, Vol. 55 (1965), 549–77, and Fred Kniffen and Henry Glassie, "Building in Wood in the Eastern United States: A Time-Place Perspective," *Geographical Review,* Vol. 56 (1966), 40–66. An ideal model of the manner in which the manmade structures of a particular area should be studied is R. W. Brunskill, *Vernacular Architecture of the Lake Counties: A Field Handbook* (London: Faber & Faber, 1974).

nucleation, and many ideas have been advanced to explain the existence of such patterns:

1. villages are found only on plains, and dispersal is characteristic of rougher, dissected, hilly areas;
2. villages are necessary where sources of water are localized, and dispersal is possible only where water may be obtained almost anywhere;
3. villages are found in agricultural areas, and dispersal is associated with pastoral economies, where holdings must be large and the owner wishes to be near his animals;
4. villages have been established for protection and defense in frontier areas and contact zones;
5. villages were founded when the land was settled under the control of some central authority such as the government, the church, a company, or even a single colonizing landlord;
6. villages were the original settlements, on the better soils, and dispersal represents a later colonization of the poorer areas by landless workers from the village;
7. dispersal from villages was a product of the enclosure of the open fields, when the farmland was reorganized into solid blocks upon which each farmer could build his own farmstead.

Although this formidable list certainly is not exhaustive, it does provide some notion of the large number of different, sometimes even contradictory, hypotheses which have been put forward to account for differences in the nucleation and dispersal of farmsteads, but a complex settlement pattern seldom has any one single, simple, satisfactory explanation. Furthermore, nucleation and dispersal may not be quite as distinctive as some scholars have believed, because an isolated farmstead may become the focus for a nucleated settlement. One example would be the "quarters" in which slaves were housed on cotton plantations in the American South before the Civil War. Another would be the "agricultural villages" of housing units for migrant workers who come to specialty farming areas in the United States to plant and harvest the crops. And in parts of Britain farmers have provided rent-free "tied cottages" near their farmhouses for their workers; the cottage is "tied" to the job, and the farm worker who quits his job must also leave his cottage.

THE MORPHOLOGY OF VILLAGES.   Scholars in Germany have done an enormous amount of research on the form of agricultural villages since 1895, when August Meitzen published his massive volumes on the settlements and agriculture of the west and east Germans, Celts, Romans, Finns, and Slavs.[7] Meitzen's ideas were closely related to the ethnic theories which were then popular. He suggested that contemporary patterns of farmstead distribution were a fairly faithful indicator of the cultural group which had first permanently settled an agricultural area: Celtic

---

[7] August Meitzen, *Siedelung und Agrarwesen der Westgermanen und Ostgermanen, der Kelten, Römer, Finnen, und Slaven* (Berlin: Wilhelm Hertz, 1895).

peoples in isolated farmsteads, Slavs in round and street villages, and the superior Germans in irregular village clusters.

Since Meitzen's time German scholars have recognized the unpopularity of the ethnic theories, and they seem to be generally "agreed that changes in subsequent economic and social history are of far greater importance," insofar as the ground plan of a village is concerned, than the language which was spoken by its original settlers.[8] Perhaps some are a bit more convincing than others; Gutkind, for example, said that "wherever Frankish influence was at work, the street village was the characteristic form of settlement, (but) . . . this does not mean that the street village was the ethnological form of the Frankish race specifically." [9]

Most German scholars, after discharging their obligation to make a disclaimer about ethnic origins, agree that the old German lands west of the Elbe and Saale Rivers, where the Germanic tribes first settled, have village layouts quite different from those in the newer lands to the east, which were occupied by Slavs until German settlers colonized them during the Middle Ages. The irregular clustered village *(Haufendorf)* is characteristic of the areas of older German settlement, whereas the round village *(Rundling)*, the line or street village *(Strassendorf)*, and the green village *(Angerdorf)* are more common in the areas of colonization to the east. Line villages are also associated with the reclamation of marshland *(Marschhufendorf)* and with the clearance and settlement of woodland areas *(Waldhufendorf)*.[10]

A brief summary cannot do justice to the Teutonic thoroughness with which German villages and their associated field patterns have been classified, but perhaps justice is impossible; discussions of village morphology in Germany can become heavily value-loaded and subjective almost to the point of mysticism. Distinctions between village types, especially between types which are quite similar, are sometimes based upon differences so extremely subtle that the classification of individual villages becomes more an act of faith than an objective exercise. This is not objectionable in and of itself, but it does pose problems when it is clouded by overtones of nationalism.

The grandiloquent terminology which has been used in schemes for the classification of village morphology in Germany has been transferred "whole-hog" to other parts of the world, sometimes with unfortunate results.[11] Perhaps this terminology has been useful, but at times one

---

[8] Gottfried Preifer, "The Quality of Peasant Living in Central Europe," in William L. Thomas, ed., *Man's Role in Changing the Face of the Earth* (Chicago: University of Chicago Press, 1956), p. 258.

[9] E. A. Gutkind, *Urban Development in Central Europe,* Volume I of The International History of City Development (New York: The Free Press of Glencoe, 1964), p. 108.

[10] Robert E. Dickinson, "Rural Settlements in the German Lands," *Annals,* Association of American Geographers, Vol. 39 (1949), 239–63.

[11] Robert Burnett Hall, "Some Rural Settlement Forms in Japan," *Geographical Review,* Vol. 21 (1931), 92–123; and *Rural Settlement Patterns in the United States as Illustrated by One Hundred Topographic Quadrangle Maps,* Publication 380 (Washington, D.C.: National Academy of Sciences-National Research Council, 1956), pp. 9, 12, and 31.

wonders whether its use was not merely a matter of intoxication with terminology. A solemn discussion of strassendorfs and haufendorfs in Japan or the United States, for example, may be useful, it may be amusing, or it may simply be sad, especially if the relationship of village form to other significant variables has not been demonstrated. "The variety of plans among the villages of England . . . is profoundly interesting," said W. G. Hoskins, but "even if we are sure of the original shape of the village, we are not yet in a position to say . . . what the various shapes and plans mean." [12]

Markets and Fairs

Many villages serve as market centers for the surrounding countryside, and the student of the rural landscape must understand the commercial functions of villages, as well as their morphology. In the New World most settlements were founded in the hope and expectation that they would become centers of commerce, but in areas of simple subsistence economy (such as medieval Europe, and much of the world even today) the great majority of settlements were originally agricultural villages, where peasant farmers lived, and the commercial functions have evolved only gradually in those which have become market centers.

People who have a subsistence economy must devote the greater part of their energies to the struggle to keep themselves alive, but even the most completely subsistent economy will produce some surpluses which its people may exchange for the surpluses produced by others. These surpluses are not large enough to support the kinds of permanent stores and market centers which are characteristic of highly specialized commercial economies, but every society, no matter how simple its economy, has some system of marketplaces where its folk foregather periodically to exchange their surpluses at markets and fairs.[13] Such marketplaces, like football stadia, may stand empty most of the time, and become crowded only during brief periods of activity.

MARKETS. The simplest technique for exchanging surpluses is bartering, which is still practiced in some parts of the world, although in modern societies the physical exchange of goods has been almost completely replaced by the exchange of money. Even in the United States, however, many a boy who grew up on a farm can remember going to the crossroads store with his mother and watching her trade a basket of eggs for goods which the family could not produce on the farm.

The best place to make a good swap is where the largest number of potential swappers are gathered together, and in pre-Norman England this meant the churchyard on a Sunday morning. Proximity to the church

[12] W. G. Hoskins, *The Making of the English Landscape* (London: Hodder and Stoughton, 1955), p. 49.
[13] J. E. Spencer, "The Szechwan Village Fair," *Economic Geography*, Vol. 16 (1940), 48–58; and Marvin W. Mikesell, "The Role of Tribal Markets in Morocco," *Geographical Review*, Vol. 48 (1958), 494–511.

also may have given the swappers a feeling, perhaps unwarranted, of greater safety from thieves and sharp dealers. In other parts of the world, as well as in early England, the congregation of the faithful, plus an aura of sanctity which helped to maintain the peace necessary for orderly business transactions, has conferred special advantages on holy places, such as churches or shrines much visited by pilgrims, as sites for markets and fairs.

The association between religion and commerce has not always been a source of comfort and satisfaction to the church, however, because markets can become rowdy affairs. "There is a widespread belief, by no means confined to Ireland, that fair days are always rainy days," said Estyn Evans, "as a consequence of the lies and profanity, the fights and pagan ways of the fair," and certainly the traditional activities at Donnybrook Fair, to cite an extreme example, were hardly conducive to quiet meditation and devotion.[14] Modern "blue laws," which are motivated by a pious desire to maintain the dignity and propriety of the Sabbath against commercial intrusion, have antecedents more than a thousand years old. Various English monarchs, beginning as early as 920 A.D., have forbidden the holding of markets on Sunday, and in 1285 Edward I even felt compelled to issue an edict against holding them in church-yards, but "it is seldom safe to regard medieval legislation as an index to anything more than the intentions of the legislator."[15]

The casual gathering beside the church remained inchoate in many villages, but in some it became more formalized, with the same set market day each week. Itinerant merchants began to make it one of the regular stops on their circuits, and in time the village burgeoned into a market town. As the market prospered and came to the attention of the Crown, it was required to pay a tax for the privilege of holding its weekly sessions, but in return it received a royal charter granting the right to monopolize the trade of the surrounding area. The holders of market charters, desirous of preventing encroachment upon their monopolies, secured legal agreement to the principle that no other charter would be granted within a minimal distance of six and two-thirds miles, or one-third of a reasonable day's journey of 20 miles. This permitted even the most distant person to reach the market in the first third of the day, spend the middle third doing his business, and still have time to return home before nightfall. Despite this legal restriction, however, many markets in the richer parts of East Anglia were less than four miles apart, and the same was probably true in other prosperous parts of England.[16]

In England the earliest markets were merely large open spaces where merchants might erect temporary booths or stalls, but in time

[14] E. Estyn Evans, *Irish Folk Ways* (London: Routledge & Kegan Paul, 1957), p. 254.

[15] E. Lipson, *An Introduction to the Economic History of England,* Vol. I, The Middle Ages (London: A. C. Black, 1920), pp. 204–8.

[16] Robert E. Dickinson, "The Distribution and Functions of the Smaller Urban Settlements of East Anglia," *Geography,* Vol. 17 (1932), 19–31.

these were replaced by more permanent structures, and the modern marketplace is lined with stores. The open-air market has not yet disappeared, however, either in England or in other parts of Europe, and traders still set up their temporary stalls in the open marketplace on a regular "market day" each week, exactly in the medieval fashion. These stalls, and their customers, can clog traffic so badly that the day of the weekly market is one of the vital bits of information about a town which is published in the annual *Members Handbook* of the British Automobile Association. Open-air markets also continue to thrive in the larger cities of Europe, as well as in the smaller country towns; in Paris, for example, the green vegetables you buy in the open-air markets are as fresh as, if not fresher than, those on sale in the green-grocery, and it is much more fun to shop in the market.

One of the more striking features in many marketplaces is the market cross, at once a reminder of the church under whose protective shadow the market was founded, an emblem of the royal authority by which the charter was granted, and a symbol of "the peace of the market," whose enforcement was required by the terms of the charter. The market cross might be no more than a simple stone spire rising from a low dais, which provided a convenient platform for reading proclamations and the like, but some were elaborate vaulted structures with an open space beneath where people could take shelter from the weather; in Chichester, for example, the large and ornate market cross at the intersection of the two main streets is an impressive structure, but in the age of the automobile it is a positive menace to traffic. Perhaps the best known market cross, one which is commonly associated with "a fine lady upon a white horse," is in the Oxfordshire village of Banbury. Banbury Cross stands at the end of a large open space which is still called the Horsefair, despite the fact that it is now a public car park.

FAIRS.   A market was commonly held on the same day each week, and attracted mainly local people. A fair was held less frequently, perhaps once a quarter or once a year, but it lasted for several days and attracted traders from greater distances. Products which ripened or became available only at certain seasons, such as wool or wheat, were sold at the end of the season in a great fair, whereas routine commodities, especially perishables, were exchanged at the regular weekly market. Many fairs, like many markets, originated in association with religious observances, and a fair was often held during the festive period dedicated to a particular saint. One of the responsibilities of wise and holy men in preliterate society was keeping the calendar, and the festivals of specific saints served a very practical purpose of reminding people about important dates in the annual cycle of agricultural production; in Ireland, for example, St. Patrick's Day marks the end of winter and the time to start the year's work in the fields.[17] An end-of-harvest festival was obviously a good time for a fair, because it assembled both the people and the surplus of their harvest.

[17] Evans, *Irish Folk Ways*, footnote 14, p. 270.

Allix has drawn a distinction between livestock fairs, in which local country people still get together at regular intervals to trade their surplus animals, and the great medieval commodity fairs, which were held at neutral frontier sites on the routeways connecting different kinds of areas.[18] The commodity fairs provided the sole mechanism for large-scale interregional exchange when long-range transportation was both difficult and dangerous. Between the twelfth and the fourteenth centuries they flourished in Flanders and Champagne, and then gradually migrated eastward as transportation facilities in Western Europe were improved and made more secure. The American agricultural fair is similar to European fairs in name only, because adequate facilities for exchange already existed at the time when it was founded, and its principal objective has been to educate local farmers in the latest scientific advances in agriculture.[19]

Villages as Central Places

CENTRAL PLACE THEORY. Unlike a subsistence economy, which can get along quite nicely with periodic markets and fairs, a highly specialized modern commercial economy requires a complex network of central places for the efficient exchange of goods and rendering of services. In recent years geographers, and especially urban geographers, have devoted a considerable amount of effort to the development of a body of theory which can describe the size, function, and locational pattern of central places.[20] This theory postulates that central places fall into a natural hierarchy (hamlets, villages, towns, cities, metropolises), and that larger places have more people, more different kinds of business activity, more business establishments, larger trade areas, and are spaced farther apart than smaller places.

Business establishments in the smaller central places provide "convenience goods" which are needed frequently, perhaps even daily, and these places must be spaced fairly closely together so that customers may obtain their goods and services with a minimum amount of travel. The larger central places, which are spaced farther apart and serve larger trade areas, offer a greater range and variety of goods and services, including "shopping goods" such as furniture and automobiles. These "big ticket" items are bought less frequently, and they justify longer trips and more serious thought before they are purchased.

It is exceedingly difficult to make any kind of general statement about the specific ensemble of goods and services which one might ex-

[18] Andre Allix, "The Geography of Fairs: Illustrated by Old-World Examples," *Geographical Review*, Vol. 12 (1922), 532–69.
[19] Fred Kniffen, "The American Agricultural Fair: The Pattern," *Annals*, Association of American Geographers, Vol. 39 (1949), 264–82; and *idem*, "The American Agricultural Fair: Time and Place," *Annals*, Association of American Geographers, Vol. 41 (1951), 42–57.
[20] Brian J. L. Berry and Allen Pred, *Central Place Studies: A Bibliography of Theory and Applications*, Bibliography Series, Number One (Philadelphia: Regional Science Research Institute, 1961).

pect to find in a central place at any given level of the hierarchy, because individual functions are not restricted to places of a specified minimal size. Furthermore, these functions reflect the needs and wishes of the people of a given region at a given time. For example, one does not expect to find grain elevators in Maine, cotton gins in Montana, or creameries in Mississippi; taverns are common in the smallest central places in Wisconsin, where drinking beer is one of the favorite indoor pastimes, but they are rare in southern and eastern Kentucky, where the majority of the people believe that beer drinking is sinful, and the sale of beer is prohibited by law.

Nevertheless, it is possible, by drawing on the work of such scholars as Trewartha, Brush, Chittick, Stafford, Salisbury and Rushton, and Berry, plus personal observation, to make a few overgeneralizations about the size of the population and the kinds of functions associated with the smallest central places, which are most intimately related to the surrounding countryside.[21] These statements apply only to the American Middle West in the mid-1960s, for significant differences have been reported from England, from the state of Washington, and from New Zealand.[22]

Hamlets, at the bottom of the hierarchy, have a population of around 100 people. A hamlet commonly has a grocery or general store, a filling station, a cafe or tavern, an elementary school, a community church, and perhaps a post office and combination feed store/lumber yard/coal yard. Small villages, one step above hamlets, have a population of around 500 people. A small village provides the same goods and services as a hamlet, and in addition it has a hardware/appliance store, a barber, a beauty shop, an implement dealer, a garage, and perhaps a drugstore and an undertaker. A large village, with a population of around 1,000 people, adds a furniture store, a clothing store, an automobile dealer, a doctor, a dentist, a bank, a high school, and perhaps a hotel or motel, a movie house, and a weekly newspaper, but it does not have a florist, a liquor store, a jewelry store, a department store, a public library, or a hospital.

[21] Glenn T. Trewartha, "The Unincorporated Hamlet: One Element of the American Settlement Fabric," *Annals,* Association of American Geographers, Vol. 33 (1943), 32–81; John E. Brush, "The Hierarchy of Central Places in Southwestern Wisconsin," *Geographical Review,* Vol. 43 (1953), 380–402; Douglas Chittick, *Growth and Decline of South Dakota Trade Centers, 1901–51,* Bulletin 448 (Brookings, S. Dak.: South Dakota Agricultural Experiment Station, 1955); Howard A. Stafford, Jr., "The Functional Bases of Small Towns," *Economic Geography,* Vol. 39 (1963), 165–75; Neil E. Salisbury and Gerard Rushton, *Growth and Decline of Iowa Villages: A Pilot Study* (Iowa City: University of Iowa Department of Geography, 1963); and Brian J. L. Berry, *Geography of Market Centers and Retail Distribution* (Englewood Cliffs, N.J.: Prentice-Hall, 1967).
[22] Robert E. Dickinson, *City Region and Regionalism: A Geographical Contribution to Human Ecology* (New York: Oxford University Press, 1947), pp. 76–92; H. E. Bracey, "English Central Villages: Identification, Distribution and Functions," *Proceedings of the IGU Symposium in Urban Geography, Lund, 1960,* Lund Studies in Geography, Series B, No. 24 (Lund, Sweden: University of Lund Department of Geography, 1960), pp. 169–90; Brian J. L. Berry and William L. Garrison, "Functional Bases of the Central Place Hierarchy," *Economic Geography,* Vol. 34 (1958), 145–54; and Leslie J. King, "The Functional Role of Small Towns in Canterbury," *Proceedings of the Third New Zealand Geography Congress* (1961), pp. 138–47.

THE DYING VILLAGE. Most contemporary Americans have never been inside a blacksmith shop or a livery stable, although nearly every village had such establishments at the turn of the century. The replacement of the horse by the internal combustion engine is related to the declining economic role of small central places in three ways: (1) some former central place functions are technologically obsolete, and have simply disappeared; (2) the tractor enables one man to handle a larger acreage of farmland, thereby reducing the size of the farm population and the number of agricultural customers who patronize any given central place; and (3) the truck for marketing and the automobile for shopping have increased the mobility of rural people. As a consequence, some of the former functions of the smallest central places have completely disappeared, and others have been lost to larger centers.

Two examples, both medical, help to illustrate the plight of the small central place. In 1968 weatherbeaten, but still legible, signs at both entrances to North Bonneville, Washington (population 494), forlornly proclaimed "We need a doctor." On March 23, 1969, the *New York Times* published a map of the 81 towns (with a median population of 1,416 persons) in upstate New York which had advertised for general practitioners at the Placement Bureau of the New York State Medical Society, and reported that many have been without a doctor of any kind for several years. The exodus of other functions has been so severe that education has assumed an exaggerated importance in the employment structure of many small places. In 1959 the Federal Reserve Bank of Chicago estimated that school teachers, janitors, and educational supervisors accounted for 17 percent of the total employed labor force in Iowa villages of less than 1,000 persons, 19 percent in villages of less than 500 persons, and 23 percent in villages of less than 250 persons.

It is sometimes assumed that the small central places which serve rural areas are losing population because they are economically moribund.[23] Studies of central place systems have concluded that the economic function of a central place is related to the size of its population, and it is perfectly logical to assume that a decline in economic function ought to be accompanied by a decline in population. But no matter how logical this may be in theory, it is not supported by the facts; between 1950 and 1960 the population of many small central places in the Middle West did not decline, but actually increased, and most of those which lost population either were in coal mining areas or had never attained a size as large as 250 persons (Fig. 10–1).[24] The total number of trade centers in South Dakota dropped from 759 in 1911 to 545 in 1951, but the number with a population of 50 or more persons increased from 382 to 397 in the same period.[25]

The larger central places of the contemporary Middle West were founded during the initial phase of settlement, and they have been grow-

[23] Berry, *Geography of Market Centers*, footnote 21, p. 115.
[24] John Fraser Hart and Neil E. Salisbury, "Population Change in Middle Western Villages: A Statistical Approach," *Annals*, Association of American Geographers, Vol. 55 (1965), 140–60.
[25] Chittick, *Growth and Decline of South Dakota Trade Centers*, footnote 21, p. 16.

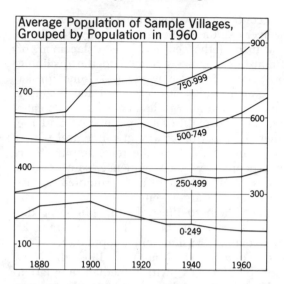

Average Population of Sample Villages, Grouped by Population in 1960

FIG. 10–1. *Average population of a sample of 400 villages in the Middle West, 1870–1970; the villages were grouped by the size of their population in 1960.*

ing ever since. Most of the smaller places which fill the gaps between them had been founded by the turn of the century, and they too have been growing. In fact, experience to date in the Middle West suggests that a trade center, once it has attained a certain minimal population (whether it be 250, 500, or some other magic number), cannot stop growing. The rate of growth may be appreciably less than the national average, and indeed, in many places it is so slow that a cynic might prefer to describe it as "upward stagnation" rather than growth, but the fact remains that the number of people is increasing, not decreasing. Why?

ACCUMULATED SOCIAL CAPITAL.   In the first place, it is important to remember that people *live* (as well as make a living) in central places, and they do not tear down or destroy part of their houses because the central place has lost part of its economic function.[26] The capital invested in the houses and other facilities of a village is one of the more important conservative factors in maintaining its population despite the fact that it is losing its economic functions. Although they may lack some modern conveniences, many village houses are large, and they are astonishingly cheap. Cheap housing has attracted low-income workers with large families, who can commute to distant jobs, and the residentially footloose, who are not tied to a specific place of residence: the retired farmer, the old-age pensioner, the divorcee trying to raise a large family on skimpy

[26] John Fraser Hart, Neil E. Salisbury, and Everett G. Smith, Jr., "The Dying Village and Some Notions About Urban Growth," *Economic Geography*, Vol. 44 (1968), 343–49.

alimony or welfare payments, or the salesman who may live anywhere in his territory. Perhaps, you may think, there aren't a great many of such people, but it does not require all that many to maintain the population of a village of less than 1,000 people.

The conservative effect of available facilities can work both ways. A few years ago, when I was doing fieldwork in an Iowa hamlet with a 1960 population of 92 souls, I was a bit unhappy to run into a string of "nobody at home" houses on one street, until I discovered that the reason was a lively bridge party in the house at the far end. The good lady who owned the house, a substantial story and a half structure, told me that she had been "taken" when she bought it. "Those highway robbers made me pay two hundred dollars," she said, "and if I had only held out a little bit longer I am sure they would have come down to a hundred and fifty!" The "girls" in the bridge game, all widows in their seventies or youthful eighties, were unanimous in their unhappiness with Mr. Jones, the grocer who operated the only business remaining in the hamlet, but nevertheless they all felt compelled to patronize his store. "Of course we younger folks could all shop someplace else," one of them told me, "but if we did and Mr. Jones had to close down, where could all of the old people who live here buy their groceries?" The quiet desperation of the situation emerged from my interview with Mr. Jones, who was eager to liquidate his capital by selling his store and stock, "but who," he asked me, "would be foolish enough to pay good money to buy a grocery store in a hick town in Iowa?" [27]

The villages of the Middle West do not yet appear to have enjoyed any significant amount of industrial migration from metropolitan areas, but this was a major factor in the rejuvenation of small towns and rural areas in the Third (Philadelphia) Federal Reserve District during much of the 1960s.[28] The new superhighway system made rural areas more accessible to metropolitan centers, declining employment in agriculture and in mining released rural labor at low wage levels in a period when the urban labor market was tight, and the cost of rural operations was reduced still further by lower land prices and tax rates.

THE DISPERSED CITY. The dispersed city hypothesis explains the persistence of many villages by postulating that they are not as moribund economically as is commonly supposed.[29] Quantitatively oriented students have assiduously counted the *number* of establishments and functions in central places, but they have failed to take notice of their *quality*, which is vastly more important in a small place with a few businesses than in a large place with many. Although the number of functions in many villages has declined since horse and buggy days, and the modern village

[27] The same idea is expressed in rather more technical language in Gerard Rushton, "Temporal Changes in Space Preference Structures," *Proceedings*, Association of American Geographers, Vol. 1 (1969), 129–32.
[28] Anne M. Clancy, "The Beginning of a Comeback," *Business Review*, Federal Reserve Bank of Philadelphia, August 1969, pp. 3–9.
[29] Ian Burton, "A Restatement of the Dispersed City Hypothesis," *Annals*, Association of American Geographers, Vol. 53 (1963), 285–89.

lacks the total "mix" of functions requisite to a complete rural service center (it would not be a village if it had them all!), at least one function has survived, and sometimes it has survived quite spectacularly.

Today many villages are dominated by a single function, and their residents drive to other villages to obtain the other goods and services they require. "For example, a man who lives in village A and works in a factory in village B will think little, on a nice spring Saturday afternoon, of driving to the new supermarket in village C to buy groceries; to the area's best hardware store, in village D, to look over a new power saw; to the automobile dealer in village E to find out whether he can wangle a better deal than he could make in Big City; and then stop off, on the way home, at the tavern in village F to have a beer with the boys in the swingingest saloon around. It beats sitting home alone watching a lousy baseball game on television." [30]

In other words, traditional farm trading centers seem to be sorting themselves out into specialized centers dominated by one activity, or only a few, but there is no reason why this should be considered unusual. A few generations ago most city dwellers lived very close to, if not actually in, the structures in which they did their work, but the two functions of residence and work became separated as cities grew in size and complexity, and as transportation improved. The various functional areas of a modern city are highly localized, and often they are separated from each other by some little distance. Private homes are uncommon in shopping centers, factories are rare in the CBD, department stores are not in industrial districts, and neither stores nor factories are welcomed as neighbors in the better residential areas. Each functional area serves the entire city, or some large portion of it, and not just those people who happen to live near by. City dwellers expect to live, work, and shop in different places, and they spend a fair portion of their time travelling from one functional area to another. Why should anyone expect dwellers in the country to be any different?

The answer seems to be that urban-oriented students of small central places are afflicted by a kind of metropolitan myopia, which makes them act as though they think that rural residents somehow are just not quite as smart and clever as city people. Berry, for example, claimed that "on one side of a breaking-point the farmers all (*sic*) travel in one direction; on the other side they travel to the other. . . . Within metropolitan regions, however, there is no such thing as an absolute breaking-point," apparently because urban dwellers are less ovine than farmers.[31] Hogwash! If Berry had only realized that the clusters of villages which comprise dispersed cities already have "an interdependent economic structure based on local specialization" (which he expects metropolitan regions to develop on a national scale), perhaps he might have spared himself the entertaining, but completely unnecessary, mental gymnastics of postulating a "phase shift" in the systematic relationships between central

[30] Hart, Salisbury, and Smith, "The Dying Village," footnote 26, pp. 346–47.
[31] Berry, *Geography of Market Centers*, footnote 21, p. 41.

places, trade areas, population served, level in the hierarchy, and population densities.[32]

The Blending of Rural and Urban

Urban ignorance of rural areas raises the question, which has been avoided thus far, of how "rural" ought to be defined. The official definition used by the U.S. Bureau of the Census is essentially negative: a place is rural if it is not urban. With some exceptions and qualifications, a place is considered urban if it has a population of at least 2,500 persons. (The U.S. Farmers Home Administration, however, sets the dividing line between urban and rural at 5,500 persons, rather than 2,500, when it makes loans on rural housing.) An official distinction between urban and rural was first made in the statistical atlas accompanying the Census of 1870, when the dividing line was 8,000 persons; this was reduced to 4,000 in 1880, and then to 1,000 in 1890, but raised back to 4,000 in 1900. For reasons which are still obscure, the figure of 2,500 was first used in a census report in 1906, and it caught on so well that it has been used at each census since then, despite the fact that no one seems to know why it was chosen in the first place.[33]

Rural people, before the First World War, were primarily farmers and their families. They lived and worked in the country, and they differed from city people in their whole rhythm and style of life, whether in dress, in speech, in manners, in diet, in their entertainment, in their politics, in their religion, or in their basic values. These distinctive and characteristic traits were so tightly interrelated that any one might be used as a surrogate for all, and the stereotypes of "city slicker" and "country bumpkin" probably had considerable justification.

These stereotypes began to break down after the First World War, when the automobile fostered greater mobility and social intercourse. As he began to participate in the social, economic, and cultural life of the city, the farmer found it useful to take on some of the protective coloration of the city dweller, and a successful modern farmer is more cosmopolitan than most city people, from whom he differs primarily in his occupation and his place of work. The automobile also took city dwellers to the countryside, first to visit, and later to purchase plots of land on which to build their homes, until eventually the residences of city people became scattered far and wide across the countryside.

By 1930 it had become desirable for the Census Bureau to make a distinction between the "rural farm" people, who lived on farms, and the "rural nonfarm" people, who did not. Although some of the rural nonfarm people live in the open countryside, most of them live in places of less than 2,500 people (which are too small to be classified as urban), or on the outskirts, but beyond the corporate limits, of larger urban centers. The

[32] Berry, *Geography of Market Centers*, footnote 21, pp. 118 and 57–58.
[33] Leon E. Truesdell, *The Development of the Urban-Rural Classification in the United States: 1874 to 1949*, Current Population Reports: Population Characteristics, Series P-23, No. 1 (Washington, D.C.: Bureau of the Census, 1949).

latter group has been increasing so steadily that in 1950 the Census Bureau introduced the concept of the "urban fringe," which is the densely populated rural nonfarm area just beyond the city limits. The people who live in the urban fringe are now classified as urban, but urban fringe areas are not defined for cities of less than 50,000 people, and a remarkably large proportion of the people who are officially classified as rural actually live on the fringes of smaller urban centers.[34]

In short, the distinctions between urban and rural have become increasingly blurred. There are still important differences between urban and rural people, but these differences are no longer tightly interrelated, and no one characteristic can be used as a surrogate for all the others. The scholar who seeks one single, simple index of urbanism, or of rurality, is doomed to disappointment, because the association between the complex of traits which formerly distinguished rural from urban has completely disintegrated, and individual traits can no longer be used as indices of others.

[34] John Fraser Hart, "The Rural Nonfarm Population of Indiana," *Proceedings,* Indiana Academy of Science, Vol. 65 (1956), 174–79.

CHAPTER **11** *mining, forestry,*
and recreation

When the countryside is taken as a whole, farming is by far the most significant employer of people, user of land, and modifier of the landscape, and thus I have accorded it the lion's share of this volume, despite the fact that nonagricultural activities such as mining, forestry, and recreation are much more important than farming in some rural areas. I have treated these activities with much greater brevity and selectivity, however, in goodly part because the geographical literature dealing with them is far less rich than that which deals with agriculture.

Mining and Forestry

The best-known recipe for rabbit stew begins with an admonition: "First catch the rabbit!" Mining and forestry are much more localized than farming, and any discussion of their impact upon the look of the land must begin be defining and identifying the areas in which these activities are important. The total value and/or quantity of production can be a misleading index of the overall importance of mining, because gold is measured by the ounce and diamonds by the carat, but coal and iron ore by the ton. The best readily available information upon which to base a map of mining areas in the United States seemed to be the number of persons employed in mining in each county, as given in the *Census of Population.*

The number of persons employed in mining varies greatly from county to county, and so does the percentage of the labor force employed in mining; these variations, unfortunately, may be totally unrelated, and neither the total number nor the percentage, if used alone, will produce a satisfactory map of mining areas. A map which shows the total number of anything by counties may be totally misleading unless it uses graduated circles, because it may show little more than variations in the size of counties.[1] Every right-thinking geographer knows that his data must be "standardized" to adjust for variations in county size, and his maps should show the number of things per square mile, not simply the total number of things.

A density map, which shows the number of mining employees per square mile, can also be misleading, however, because the census figure includes all persons employed by mining companies, administrative and clerical types as well as production workers; it does not distinguish a sweet young thing in a miniskirt chewing gum and pounding a typewriter in the head office from a grizzled old-timer in a hard hat chewing tobacco and pounding a pneumatic hammer at the face of the ore body. The density map overemphasizes the cities which have the offices of mining companies. Such an urban bias can be corrected by showing the number of mining employees as a percentage of the total labor force, but the percentage map overemphasizes counties populated mainly by miners and jack rabbits. The counties with the highest percentages are those which have precious little economic activity except mining, and very few people, including only small numbers of miners.

The best features of the density map, which shows number per square mile, and the intensity map, which shows percentage, may be combined by using the technique of minimal percentages.[2] The minimal percentage is the percentage of the total national labor force which is employed in a specific kind of activity; in the United States in 1960, for example, 1.01 percent of the labor force was employed in mining, and 0.14 percent was employed in forestry and fisheries. An activity which employs less than the minimal percentage of a county's labor force probably is not important in that county. An index of employment in any industry in any county may be calculated by subtracting the minimal percentage of that county's labor force (the number of persons who *should* be employed in the industry) from the number who actually are employed in the industry. The excess, or surplus number of employees in that industry in that county, is divided by the area of the county to standardize for area.

To simplify calculations, I rounded off the minimal percentage for mining in the United States in 1960 to an even 1 percent, and calculated an index of mining employment (I) for each county by subtracting 1 percent of the county's total employed labor force (−0.01 T) from the actual number of persons employed in mining (M), and dividing the result by the area (A) of the county in thousands of square miles (Fig. 11–1). In 1960, for example, Carbon County, Pennsylvania, had an area (A) of 405 square miles and a total employed labor force (T)

---

[1] This and other common cartographic blunders are treated in some detail in John Fraser Hart, "A Map of Mining Employment in the United States in 1960," *The Minnesota Geographer,* Vol. 22, No. 2 (April 1970), 1–8.

[2] Although the concept of minimal percentages is quite similar to the concept of ubiquity which is used in Richard Hartshorne, "A New Map of the Manufacturing Belt of North America," *Economic Geography,* Vol. 12 (1936), 45–53, I have great difficulty in thinking of mining as ubiquitous when not a single miner was employed in half the counties of the United States in 1960.

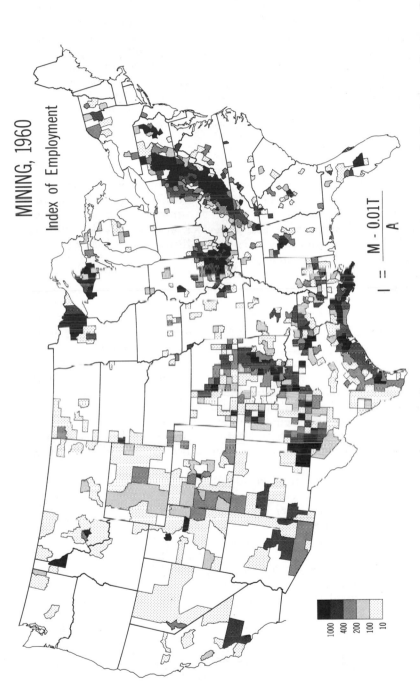

MINING, 1960

Index of Employment

$$I = \frac{M - 0.01T}{A}$$

1000
400
200
100
10

FIG. 11–1. Mining districts of the United States in 1960. The index of mining employment (I) was calculated for each county by subtracting 1 percent of the total labor force (−0.01 T) from the number of persons employed in mining (M) and dividing the result by the area of the county (A) in thousands of square miles.

of 18,590 persons, of whom 306 were employed in mining (M); subtracting 185.9 (0.01 T) from 306 (M) gave a surplus of 120.1 persons employed in mining, and dividing this surplus by 0.405 (A) produced an index of mining employment of 297 (I) for Carbon County in 1960. The mining districts which stand out quite clearly are the coal fields of Appalachia and the Eastern Interior; the oil fields of Texas, Oklahoma, Louisiana, and adjacent states; the iron ore and copper mining areas of the upper Lake states; and the mineralized areas of the Mountain states.

Mining Camps

Nearly all mining camps are temporary, and most of them look the part. The name itself, "camp" rather than "town," implies their impermanence. A sea of tents and shacks springs up almost overnight when a mineral deposit is discovered, and men flock in to seek their fortunes; there are few jobs for women, and none at all for ladies. A few mining camps have been fortunate enough to attract alternative means of support, and struggle on after the mines have closed, but most have been created with the single-minded purpose of extracting a mineral from the earth, and the camp folds up when the mineral resource has been exhausted. The entire mining landscape is distinctive: devastated hillsides, abandoned workings, ugly waste dumps, broken and rotting machinery, dreary rows of tired and weatherbeaten houses, ghost towns, rutted roads, and rusting railroad tracks.[3]

Mineral deposits often are discovered in remote and isolated places where no one in his right mind would go on a bet: the icy wastelands of northern Alaska, the searing deserts of the Middle East and Africa, the precipitous chasms of the mountain West, the swamps and marshes of Texas and Louisiana, the hills of Kentucky and West Virginia. A network of roads, railroads, canals, and pipelines must be developed to bring in the workers, and their tools and equipment, and to carry out the mineral once it has been removed from the ground.

The actual extraction may take place at well, underground mine, or open pit. Wells are used for petroleum, natural gas, and other gases and liquids. Oil fields are dotted with tens, sometimes hundreds, of wellhead pumps, which look like giant insects as they bob up and down to lift the precious fluid to the surface. Pipes carry it to nearby clusters of tanks for temporary storage. The tall metal derricks which were used to drill the wells are a common sight in older oil fields, but the newer wells are drilled by portable rigs which can be moved to another site after the well has been completed.

The most distinctive feature of an underground shaft mine is the

---

[3] The melancholy landscapes of mining areas have attracted few studies by geographers; one of the better ones is R. T. Jackson, "Mining Settlements in Western Europe: The Landscape and the Community," in R. P. Beckinsale and J. M. Houston, eds., *Urbanization and Its Problems: Essays Presented to E. W. Gilbert* (Oxford: Basil Blackwell, 1968), pp. 143–70.

headframe which supports the hoisting sheaves, great grooved wheels which carry stout metal cables from machines in the winding house to elevator cars in the shaft. The machines unwind to lower the cars with men and machinery to the working level deep within the earth, and wind to raise cars loaded with ore, or with tired and dirty miners at the end of their working shift. Close by the winding house are the lamp house, where the men can change clothes and shower, and a nondescript clutter of repair shops and storage sheds.

Opencut mining is practical only where a mineral body is close to the surface.[4] The worthless overburden is stripped away and dumped to one side, and the exposed mineral is blasted loose, loaded into enormous trucks, and hauled to nearby processing plants to be prepared for shipment. Opencut mining has produced some of the most spectacular man-made features on the surface of the earth. The great pyramids of Egypt would be lost in the copper mine at Bingham Canyon, Utah, which is half a mile deep and two miles wide, or the iron ore mine at Hibbing, Minnesota, which is three miles long and a mile wide. Some of the world's largest earth-moving machinery is used in opencut mines, and the experience of "riding" a giant power shovel is rather like being in the engine room of a destroyer in a moderately heavy sea.

Hundreds of thousands of acres in the United States have been devastated by opencut strip coal mines.[5] In level areas a large power shovel digs a trench down to the top of the coal seam, dumps the overburden in a steep ragged ridge to one side, and continues the trench straight across country to the edge of the property.[6] Smaller shovels remove the coal from the bottom of the trench. At the end of the property the power shovel reverses direction, cuts a new trench parallel to the first one, and dumps the overburden from the second trench into the one from which the coal has just been removed. The repetition of this procedure results in roughly parallel lines of steep ragged spoil banks, which give the area a corrugated appearance. The final trench remains empty, a long deep pit with a vertical high-wall of solid rock on one side and a steep spoil bank on the other. Eventually it may fill with rainwater and become a long narrow lake, but the coal-bearing strata often have such a high sulfur content that they make the water poisonous to fish and quite unpleasant for swimmers.

Strip mines do even more damage in hilly country. The power shovel hacks a bench tens of feet wide out of the hillside and dumps the spoil downslope, where it knocks down trees and houses, buries

---

[4] Two special types of opencut operations are placer mines, which use water to sort heavier mineral matter from deposits of sand and gravel, and quarries, which produce squared dimension stone for construction or monumental purposes.

[5] John Fraser Hart, "Loss and Abandonment of Cleared Farm Land in the Eastern United States," *Annals,* Association of American Geographers, Vol. 58 (1968), 429–30.

[6] Arthur Doerr and Lee Guernsey, "Man as a Geomorphological Agent: The Example of Coal Mining," *Annals,* Association of American Geographers, Vol. 46 (1956), 197–210.

fields and gardens, dams up creeks, and destroys everything else. The bench winds around the hillsides, following the coal outcrop along the contours, and makes an ugly scar which meanders across the country-side for miles and miles.

Nearly all mining activities produce large quantities of waste material which the mine operator wants to dispose of as simply and cheaply as possible, preferably by dumping it near the place where it was produced. Mine development to obtain access to a mineral body removes considerable amounts of barren material, whether overburden from a strip mine or rock excavated in sinking a mine shaft. Many ores contain only minute quantities of the mineral for which they are mined; the Homestake Mine at Lead, South Dakota, for instance, averages slightly less than three ounces of gold for every ten tons of ore brought to the surface. As much waste material as possible should be removed at the mine to save the cost of shipping it, but crushing the ore creates dust, and washing it produces tailings which can pollute streams and lakes. Mine waste dumps are unstable, especially after they have been soaked by heavy or protracted rains, and the devastating effects of their collapse have been demonstrated time and again, as at Aberfan in Wales and along Buffalo Creek in West Virginia.

Physical waste material can be separated and dumped at the mine mouth, but most ores also contain chemical impurities which must be removed at a smelter or refinery. Chemical treatment produces enormous quantities of slag and noxious gases which may wipe out plant life for miles around, and most smelters have towering stacks to release the gases high in the atmosphere so they will be diluted as much as possible and spread over a wider area; the "big stack" at Sudbury, Ontario, is nearly a quarter of a mile high.

Uneven subsidence of the land surface may also create problems in areas where large quantities of material have been removed from underground mine workings: houses sag and crack, roads buckle unexpectedly, water accumulates in strange places, streams pond up and overflow their banks. New development is discouraged, because most people are reluctant to invest their money in factories or office buildings that may suddenly start to sink into the ground.

The distribution of houses in mining areas seems to be related to the type of mining activity. Houses are scattered all over the country-side in areas where mines are dispersed, as in oil and gas fields, and in strip mining areas where the mining operation is constantly on the move, but they are clustered in company towns near the mouths of shaft mines. The company town is dominated by the headframe, tipples, and other mine structures, and by the large building which holds the company store and offices. Unprepossessing houses line the streets, which are laid out in neat geometric patterns. Apart from a small separate area of slightly better houses for supervisory personnel, the standardized company houses are all of one class and one type in order to reduce costs; many of them will be dragged down the road to another camp when the local mine plays out and is shut down.

Forestry

I prepared a map of forestry and fishery areas in the United States in 1960 by using the same minimal percentage technique that I used in preparing the map of mining districts. The minimal percentage of 0.14 for forestry and fisheries was rounded off to two-tenths of 1 percent, and an index of employment in forestry and fisheries (I) was calculated for each county by subtracting two-tenths of 1 percent of the county's total employed labor force (-0.002 T) from the actual number of persons employed in forestry and fisheries (F), and dividing the result by the area (A) of the county in thousands of square miles (Fig. 11–2). Forestry, unfortunately, could not be separated from fisheries, because the census data do not distinguish between them at the county level. The highest values are in coastal counties, where fishermen are concentrated; the principal areas of forestry are the Southeast, the Pacific Northwest, and the upper Lake states, with lesser concentrations in the uplands of the Northeast and the southern Rocky Mountains.

PACIFIC NORTHWEST. The old logging camp, with its bunkhouse, commissary, and machine shops, has virtually disappeared in the Pacific Northwest, because modern loggers live in town and commute to their jobs in the woods each day.[7] They fell the trees in patches, leaving clumps and strips to reseed the logged-off areas and to serve as firebreaks between them. The logs are cut into 33-foot lengths for hauling to the mill by logging truck or railroad. Logs cut in distant areas are hauled to protected bays, assembled into giant "rafts," and towed to the mill through sheltered coastal waters by powerful tugboats. The mill has a pond where logs can be stored and sorted.

Large sawmills are widely scattered through the Northwest, but complete and efficient utilization of a variety of tree species also demands mills which can convert the better Douglas fir logs into plywood, and mills which can use the smaller logs and sawmill waste to make paper. A modern forest industry complex includes plywood mills, sawmills, and pulp and paper mills. At the pulp mill the logs are cooked with chemicals to separate the cellulose fibers from the lignin which holds them together, or shredded and mixed with water to form a soupy pulp. A thin mixture of pulp and water is poured onto a continuously moving wire screen, the water is drained off, and the wet mat of fibers is passed through rollers and driers, from which it emerges as a ribbon of paper. Chemical fumes and waste water from pulp and paper mills have created some of our most severe pollution problems.

---

[7] Two useful case studies of forest operations in the Pacific Northwest are Walter G. Hardwick, "Port Alberni, British Columbia, Canada: An Integrated Forest Industry in the Pacific Northwest," in Richard S. Thoman and Donald J. Patton, eds., *Focus on Geographic Activity: A Collection of Original Studies* (New York: McGraw-Hill, 1964), pp. 60–66; and "Toledo, A Forest Based Community," in Richard M. Highsmith, Jr., ed., *Case Studies in World Geography: Occupance and Economy Types* (Englewood Cliffs, N.J.: Prentice-Hall, 1961), pp. 102–10.

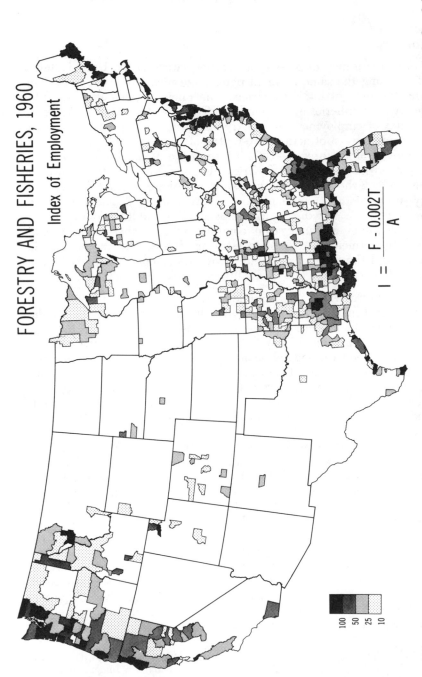

FORESTRY AND FISHERIES, 1960

Index of Employment

$$I = \frac{F - 0.002T}{A}$$

FIG. 11–2. *The index of employment in forestry and fisheries (I) for each county in the United States in 1960 was calculated by subtracting two-tenths of 1 percent of the total labor force (−0.002 T) from the number of persons employed in forestry and fisheries (F) and dividing the result by the area of the county (A) in thousands of square miles.*

SOUTHEAST. Although the Southeast is one of the most heavily wooded areas in the United States, and despite the increasing importance of forest industries in the region, the impact of forestry upon the landscape of the Southeast is a virtually unexplored aspect of the geography of the United States.[8] Part of the lack of interest in forestry can be explained by the regional tradition of brush fallow: trees were weeds that had to be cleared before land could be returned to cultivation, and "setting the woods on fire" was a popular form of outdoor recreation. Small portable sawmills have left piles of sawdust scattered through the backwoods, but the sawmill operator has ranked somewhere between the garbage collecter and the bootlegger.

The first major commercial products of the southern pine forests were "naval stores," turpentine and rosin, which are still important in southern Georgia and northern Florida.[9] The woods worker chops a V-shaped gash through the bark of each tree, and nails a metal cup at the bottom. The resinous sap which oozes into the cup is taken to a turpentine still to be refined and prepared for shipment.

The pulp and paper industry has boomed in the Southeast since World War II; in 1952 the region had 66 pulp and paper mills with an average daily capacity of 414 tons, but by 1971 the number had increased to 123 mills, and their average daily capacity was nearly 750 tons (Fig. 11–3).[10] Pulpwood is trucked to the mills from a radius of 50 to 60 miles, and woodland owners at greater distances can haul their logs to local collecting yards, whence they are carried to the mill by rail.

The pulp and paper business has expanded so rapidly that mill owners have had difficulty obtaining enough pulpwood from private landowners, and paper companies have begun buying land in order to ensure a dependable supply of raw materials. In 1970 the wood-using industries owned 67 million acres of commercial forest land, almost 3 percent of the entire United States, an area nearly as large as the two states of Illinois and Iowa combined.[11] Nearly half of this land, 31 million acres, was in ten southeastern states (Fig. 11–4). Companies in the forest industry owned one-tenth of the entire ten state area, and some counties were virtually "company counties." [12] When and how was this land acquired, from whom, and how is it managed? What has been the landscape impact of these large new landholdings?

[8] One of the rare exceptions is George A. Stokes, "Lumbering and Western Louisiana Cultural Landscapes," *Annals,* Association of American Geographers, Vol. 47 (1957), 250–66.

[9] John Fraser Hart, *The Southeastern United States,* Searchlight Book No. 34 (Princeton, N.J.: Van Nostrand Reinhold, 1967), pp. 30–35.

[10] The Southern Forest Experiment Station in New Orleans publishes an annual report on pulpwood production and pulpmill location and capacity.

[11] *Statistical Abstract of the United States, 1972,* p. 626.

[12] For example, forest industry companies owned two-thirds of Taylor County and four-fifths of Dixie in Florida's armpit; the "company county" would seem to offer fascinating possibilities for research.

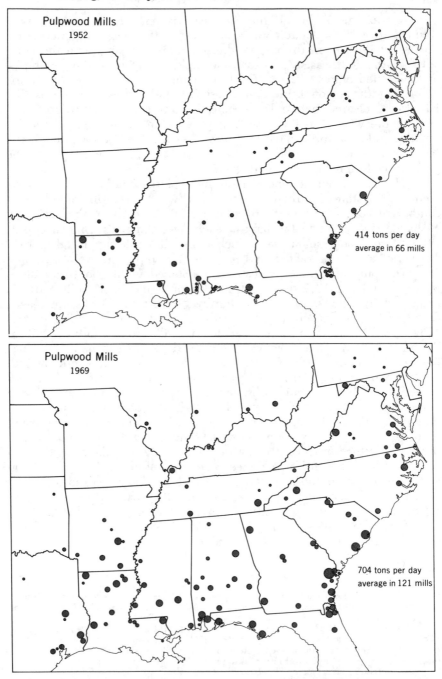

FIG. 11-3.   *The number and size of pulpwood mills in the southeastern United States has been increasing rapidly since World War II.*

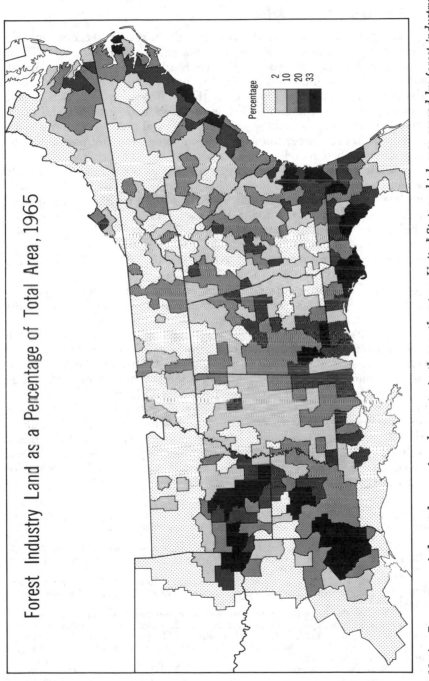

Forest Industry Land as a Percentage of Total Area, 1965

Percentage

2
10
20
33

FIG. 11–4. *Percentage of the total area of each county in the southeastern United States which was owned by forest industry companies in 1965.*

The Impact of Wealth

The greatest changes in the countryside in recent decades have resulted from the demands which city folk have placed upon it. These demands will surely increase in the years ahead. Most modern urban Americans are working shorter hours for more money, with longer weekends and longer paid vacations than their fathers enjoyed. They have increasing amounts of money and time to spend as they see fit, and many of them are electing to make greater use of rural areas. City people are fanning out in ever larger numbers over an ever widening portion of the countryside in their search for recreation.

Escape to the countryside has been beyond the means of most city folk until recently, but the idea has long been attractive to those who have been able to afford it. Rich and powerful Englishmen have been building palatial mansions in rural areas for nearly half a millennium; "Windsor and Hampton Court are monuments to the notion that royalty belongs in the country." [13] A gentlemen's "stately home" had to have an appropriately spacious setting, of course, so he fenced off a vast area around it and converted it into parkland. The park was an open stretch of grass, studded with scattered clumps of trees and decorated by flocks of cattle or sheep, which added bucolic charm while they kept the grass neatly trimmed.

Parks, like forests, originally were created to satisfy the Norman passion for hunting. William the Conqueror, when he divided up the land among his henchmen, set aside certain parts of his new kingdom as his own royal hunting preserves.[14] The royal hunting preserves were called forests, and they were subjected to special forest laws, with severe penalties for poaching and trespass.[15] It mattered not whether the area was wooded, although many forests were, and some still are.[16] Initially a park differed from a forest only because it was enclosed, but the dense woodland of the park gradually became more open as grazing animals thinned out the trees and prevented their regeneration by grazing or trampling the saplings.[17]

---

13 W. G. Hoskins, *The Making of the English Landscape* (London: Hodder and Stoughton, 1953), pp. 126–37; the quotation is from David Lowenthal and Hugh C. Prince, "English Landscape Tastes," *Geographical Review*, Vol. 55 (1965), 189.

14 Hoskins, *The Making of the English Landscape*, footnote 13, p. 73; and Philip H. Nicholls, "On the Evolution of a Forest Lansdcape," *Transactions*, Institute of British Geographers, No. 56 (1972), 57–76.

15 The most heinous crime committed by Robin Hood and his merry men was hunting the red deer in Sherwood Forest, where only the king had the privilege of hunting.

16 Parts of the New Forest have superb oak and beech woods, but Sherwood Forest proves disappointing to many visitors, and one pious worthy, after traversing Ettrick Forest in the Scottish Borders, remarked: "If Judas Iscariot had betrayed Our Lord in Ettrick Forest, he could not have committed suicide for want of a tree from which to hang himself."

17 Many relict farm woodlots in the eastern Middle West have been converted into attractive open "parkland" by the same process; cattle have browsed the branches as high as they can reach, and they have grazed or trampled the seedlings before they can become established.

It is romantic, and certainly not unreasonable, to trace the Norman passion for hunting back to the great Aryan expansion of about 1800 B.C., when nomadic horsemen from Central Asia suddenly began thundering out over much of Eurasia. The Achaeans of Greece, the Hyksos of Egypt, the Hittites of Anatolia, the Aryans of India, the Shang of China, probably all were part of this expansion, which has been bedeviled by overtones of racism because the name Aryan came to be applied to the Indo-European family of languages, and then it was assumed that these languages were related to a racial type. In fact, these were not mass invasions by destructive nomads; the number of invaders was so small that they were absorbed into the native population, and the local languages were not changed, as they would have been if there had been mass slaughter and replacement of peoples.

The Aryan invasions, like the Norman Conquest of England, were the work of a small number of skillful leaders and organizers, with a few camp followers. Despite their number, they had tremendous power, because they had learned to ride horseback, they had invented the wheel and learned to use horses to pull their fighting chariots, and quite probably, they had learned the secret of ferrous metallurgy, and could fashion their weapons and tools of iron rather than bronze. They came as conquering horsemen, taking unto themselves the best serfs and the richest lands, where they could be supported by a productive but passive peasant population while they indulged in their favorite pastimes of hunting, fighting, and riding horses.

To this day the heritage of the man on horseback, the conquering warrior, remains a remarkably strong force in European ruling life and society.[18] It generated an epic literature of chivalry and a romantic tradition of cavalry, although the horse in war disappeared quite abruptly in 1914. Trousers, which were invented by the horse-riding nomads, are still the proper garment for the lordly male, and one which many males feel should remain peculiar to their sex. Horse racing, which the nomads enjoyed, remains "the king of sports and the sport of kings," although Europeans use jockeys, whereas the Mongols used small boys as riders. And fox-hunting, riding to hounds in pursuit of the elusive fox, remains a highly prestigious activity among those who consider themselves aristocrats, or would like to be so considered.

Organized fox-hunting appears to have originated in the East Midland counties of England around 1750. Although some apologists maintain that the purpose of fox-hunting is to get rid of foxes, which are a nuisance in rural areas, nothing is better calculated to dismay an avid fox-hunter than the prospect of having no foxes to hunt.[19] Many hunting landlords in the late eighteenth and early nineteenth century took steps to forestall

[18] One of the legacies of the Aryan invasions was a strongly class-structured political system, with sharp distinctions between chevalier and villein, knight and peasant, the powerful man on horseback and the powerless man on foot.
[19] On September 29, 1962, *The Times* (London) published an article from a fox-hunter who exultantly reported that the fox "disease which ruined the past two seasons in East Midland counties has disappeared and, happily, the fox population is back to normal."

such a disaster; they created "fox coverts," sheltered breeding places for foxes, by planting spinneys and clumps of gorse in well-chosen spots on their estates. Sending a pack of hounds through a covert was almost certain to raise a fox and guarantee a good day's hunting. Fox coverts are still scattered over the East Midlands, often the only trees in sight over thousands of acres, and they are clearly labelled on the one-inch maps of the Ordnance Survey.

Fox-hunting was exported to the United States in 1877, when the first pack of hounds was established at Meadow Brook, Long Island. Some hunts in the United States pursue a dragnet scent rather than a real live fox, but the ritual and regalia are assiduously patterned after the British model: packs of wealthy suburbanites adorn themselves in scarlet jackets, mount their horses, quaff their stirrup cups, and pelt across the countryside in pursuit of a pack of yelping hounds, to the intense astonishment of innocent passersby. Fox-hunting in the United States is a phenomenon of the outer fringes of the suburbs, with the greatest concentration in the Gentleman Farm Belt just west of the nearly continuous belt of cities along the Eastern Seaboard (Fig.11–5).

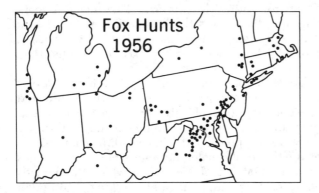

FIG. 11–5.   *Fox hunts in the northeastern United States in 1956. Except for San Francisco, each of the 12 largest cities in the United States has at least one fox hunt, but hunts are concentrated in the Gentleman Farm Belt west of the southern part of Megalopolis, from Charlottesville, Virginia, northeastward to the Hudson Valley. Compiled from the list in* Baily's Hunting Directory, 1955–1956 *(London: Vinton and Co., 1955).*

The Gentleman Farm Belt is one of the most elegant and least known agricultural regions in the United States.[20] The lush rolling countryside literally reeks of wealth and its appurtenances. Registered cattle and thoroughbred horses graze carefully manicured pastures dotted with shade trees and enclosed by handsome board fences or rugged stone walls. The fine old houses are set back from the road, discreetly screened by trees from public gaze. The lands of the estates are interspersed with

[20] This belt appears to have been ignored completely by American geographers, except for a fleeting reference in Edward Higbee, *American Agriculture: Geography, Resources, Conservation* (New York: John Wiley, 1958), pp. 306–7.

equally attractive golf courses, greenhouses and nurseries, and the neat grounds of private schools, boarding and country day. Along the roads are antique and specialty gift shops, old book stores, and good country inns and restaurants, which cater as much to tourists as to the local residents.

The Gentleman Farm Belt, which lies just west of four of the 14 largest cities in the United States, is merely the epitome of the prestige suburban area. Every American city has one, the master bedroom for the city's top brass, where city and country merge in large estates, and well-to-do people pay outlandish prices to live near other WASP, Ivy League, Republican, fashionable, conservatives from some of the best old families who belong to the best country club.

The prestige suburbs inevitably attract office buildings and factories, especially the newer kinds with low profiles that can be landscaped to blend unobjectionably into the local scene. The process begins when a company president suddenly asks himself, "Why should I have to go to the office/plant every day when I could just as easily bring the office/plant right here? Most of the other members of the top-management team also live here, or within easy commuting distance. Too bad about the poor fellow who was foolish enough to buy a house in a suburb on the other side of the city; as for the workers, we simply must remember to provide enough parking space for them." In time, of course, the roadsides become festooned with little clots and blobs of jerry-built ranch houses for the workers, wherever a developer could buy enough land to put up ten or 20.

Financial considerations as well as prestige can attract men of wealth into gentleman farming, because much of the money they spend on land and livestock is tax deductible.[21] For example, a man with an adjusted gross income of $100,000 in 1972 was in the 60 percent tax bracket; he could keep only $4,000 of each additional $10,000 he made, but he could write off the entire $10,000 as a tax deduction if he invested it in cattle. If he broke even when he sold them, he would make $10,000 on an investment which had only cost him $4,000; he could write off any loss as a tax deduction against his off-farm income; and any profit could be reinvested in the farm and become taxable at the much lower capital gains rate.

This country needs more gentleman farmers, because they can afford to take the financial risk of experimenting with new ideas and new techniques, and most working farmers cannot. Furthermore, city owners of rural land generally have a major responsibility for maintaining and even improving the quality of the rural environment, because they can afford to spend money to keep it attractive, as they want it to be, and most farmers can no longer afford to do so.

The impact of wealth upon the American countryside is nowhere more impressive than on the horse farms of the Kentucky Bluegrass and the Ocala area of north central Florida (Fig. 11–6). Both areas have rich

---

[21] A handsome color spread on executives who have sought high profits, depreciation write-offs, and capital gains advantages by acquiring ranches appeared in "Down to the Ranch on Weekends," *Fortune*, October 1968, pp. 152–59.

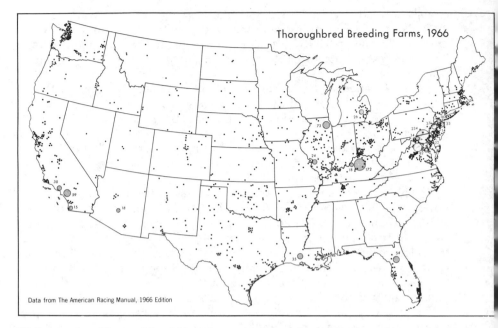

Thoroughbred Breeding Farms, 1966

Data from The American Racing Manual, 1966 Edition

FIG. 11–6. *Thoroughbred breeding farms in the United States in 1966. Compiled from addresses in the* American Racing Manual *(Chicago: Triangle Publications, 1966).*

soils derived from phosphatic limestone, rolling topography, verdant pastures with white board fences and large old shade trees, fancy entrance gates, paved farm roads, and handsome horse barns.[22] Most owners visit their farms only infrequently, and leave day-to-day affairs in the hands of skilled resident managers.[23] Both areas have the infrastructure necessary for rearing and training race horses: tack shops, blacksmiths, special feed stores, veterinarians, the choice of a large number of high quality stallions standing at stud nearby, and fully equipped training tracks.

The Kentucky Bluegrass country, which is the loveliest rural area in the United States, has a long tradition of breeding racehorses.[24] The Ocala area has milder winters, which allow the horses to graze outside throughout the year, and to train on tracks which are never frozen. Good quality land is plentiful near Ocala, and it is not nearly so expensive as

[22] Alice Luthy Tym and James R. Anderson, "Thoroughbred Horse Farming in Florida," *The Southeastern Geographer,* Vol. 7 (1967), 50–61.

[23] Many absentee horse-farm owners also have showplace farms in other parts of the country, such as quail-shooting plantations in Georgia; Merle Prunty, Jr., "The Woodland Plantation as a Contemporary Occupance Type in the South," *Geographical Review,* Vol. 53 (1963), 1–21.

[24] An outstanding guide to the area is Karl B. Raitz, ed., *A Tour of the Bluegrass Country,* Kentucky Study Series, Number Four (Lexington, Ky.: University of Kentucky Department of Geography, in cooperation with the Fayette County Geographical Society, 1971).

land in the Bluegrass; farm labor in the Ocala area is also cheap and plentiful. Despite these advantages, however, Ocala did not become a major breeding center until after locally bred horses had won the Kentucky Derby in 1956 and in 1962, but in recent years the area has boomed.

The Impact of Leisure

Before World War II the American countryside belonged to the rich and to the farmers, but since then other groups have been demanding a share of it. Increased leisure time—shorter work weeks and longer vacations—has been more important than higher income. Money has been necessary, of course; it always is. People in the lower income brackets still have inadequate access to outdoor recreational opportunities. The steadily rising demand has been generated primarily by the well-to-do and middle income groups.

The demand has not been homogeneous. Recreational preferences differ markedly by age, sex, race, place of residence, and income/educational/occupation. Professional and managerial types prefer outdoor activities which combine uplift and enrichment with peace and quiet: nature study, photography, hiking, sailing, bird watching.[25] Blue-collar people want power and blood: speedboats, snowmobiles, hunting, fishing. The mechanic riding a snowmobile has no use for the college kid on skis, and vice versa, their entire value systems are different, and they are truly people of different cultures.

The campground is the poor man's summer resort, the cheapest place to spend a family vacation. One of every five Americans went camping in 1970, and he averaged 11 days in his tent or trailer. His age was between 30 and 44 years, he lived in the suburbs, and he had a wife and two children. The big camping period was the school vacation months, June, July, and August. The family sought a campground with woods and water: a shady spot in a scenic area near a nonpolluted stream or lake where they could swim, fish, and boat. They probably found one in a national forest or a state park.[26] If they stopped at a privately developed campground they were more likely to expect a playground, sports equipment, laundry facilities, and perhaps even a swimming pool.

[25] One of the finest possible forms of outdoor recreation, as far as the student of the countryside is concerned, is visiting open-air museums of the sort which have been developed in Scandinavia and the Netherlands. Buildings typical of the various parts of the country have been assembled on a single site, and they are placed in a setting which is as close as possible to their place of origin, including the plants in the fields, the animals grazing them, and the fences around them. Pioneer settlements have been assembled or reconstructed, albeit without the regional motif, at James Fort in the Jamestown Festival Park in Virginia, Spring Mill State Park in Indiana, and Lincoln's New Salem State Park in Illinois.

[26] There seems to be no simple way of figuring out how people select campgrounds, or why they prefer one to another; Robert C. Lucas, *User Evaluation of Campgrounds on Two Michigan National Forests,* Forest Service Research Paper NC-44 (St. Paul, Minn.: North Central Forest Experiment Station, 1970), p. 8. The U.S. Forest Service provides more outdoor recreational opportunities than any other Federal agency, and it has been responsible for some of the best and most useful research on outdoor recreation.

Growing demand has encouraged the development of public and private campgrounds in recent years.[27] The average campground has 50 to 60 acres of land with 25 developed campsites consisting of a parking spur, picnic table, tent pad, fireplace, and garbage can. The private campground industry today is similar to the motel industry before World War II, when individual operations offered improvised facilities and were run by older people.

SECOND HOMES AND SPECULATORS.   In the nineteenth century only a few Americans could afford vacations. They spent their entire time at the nicest resort hotel they could afford, and enjoyed long quiet days admiring the view from the rocking chairs on the front porch. By the turn of the century wealthy magnates had begun to build palatial "summer cottages" in fashionable resort centers, such as Newport, Rhode Island, and lesser folk were buying abandoned farmhouses and fishermen's shacks for conversion into summer homes. Most of the early summer cottages richly deserved their name. They were primitive cabins, very much on the order of contemporary hunting shacks, where hearty male types could convince themselves that they were really getting back to nature. In time females began to exercise their civilizing influence, and since World War II cottages have been better built, better equipped (inside plumbing, adequate heating and electricity, but rarely a telephone), and much more expensive. Many cottages have been winterized for year-round use as genuine second homes, and their owners plan to live in them permanently after they retire.[28]

The summer cottage belt of the United States extends along the east coast from Bar Harbor, Maine, to Virginia Beach, with an extension westward along the shores of the Great Lakes into the morainic lake belts of the Upper Midwest (Fig. 11–7). The major colonies are within driving distance of large urban centers. An indeterminate number of summer cottages pass out of the second home category each year because they have been converted into year-round homes for commuters or retired people.[29]

[27] Between 1961 and 1967 the number of public campgrounds in the 11 northeastern states increased from 273 to 377, and the number of private campgrounds shot fom 160 to 1,428; George H. Moeller, *Growth of the Camping Market in the Northeast*, Forest Service Research Paper NE-202 (Upper Darby, Pa.: Northeastern Forest Forest Experiment Station, 1971), p. 1; the public campgrounds were more likely to be in sparsely populated, heavily forested, mountainous areas, but most of the growth, public and private, occurred within five miles of major water bodies.

[28] John Fraser Hart, "A Rural Retreat for Northern Negroes," *Geographical Review*, Vol. 50 (1960), 147–68.

[29] Statistics on the number of summer cottages, second homes, or vacation homes by county in the United States can only be described as a mare's-nest. The Census of Housing has published data on the number of "seasonal vacant housing units" in each county since 1940, but not all of these are summer cottages, because some are for migrant farm workers. The national total rose from 1,050,466 in 1950 to 1,742,465 in 1960, but then dropped to 1,022,464 in 1970 because some "seasonal vacant housing units" appear to have been transferred to the "held for occasional use" category; data on housing units held for occasional use have been published only for counties in metropolitan areas, although they are available on second count summary tapes for all counties. *The Statistical Abstract of the United States*, 1972,

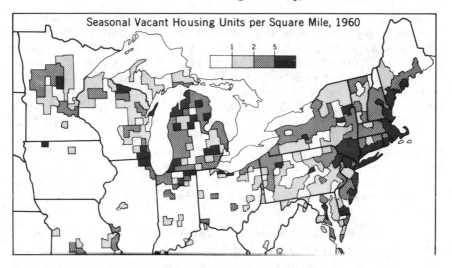

FIG. 11–7. *The number of seasonal vacant housing units in each county is a surrogate for the number of summer cottages, but an indeterminate number of housing units for migrant workers is also included.*

The very features which made an area attractive to summer cottage owners virtually ensure that it will be drastically modified, if not actually destroyed, by overcrowding, because more and more people keep flooding in to enjoy them.[30] Each cottage owner wants to keep the area precisely as it was when he first saw and was attracted to it, and he resents all who have come after him, often without realizing that he himself is equally resented by those who have been there longer than he has.[31] Although he may be quite vocal, he has little or no voice in the affairs of the community, because he is registered to vote at his permanent place of residence, and is not permitted to vote in local elections, no matter how gravely they may affect his property.

The four to six billion dollar a year demand for second homes has

---

contains the mind-boggling information that the number of American households reporting ownership of second homes almost doubled in only three years, from 1,676,000 in 1967 to 2,890,000 in 1970 (p. 684), but apparently the latter figure includes tents, boats and yachts, campers and trailers, homes outside the United States, and shared-ownership cottages which were counted more than once because they were reported by each owner. Most of this information came to me in a letter from Arthur F. Young, Chief of the Housing Division of the Bureau of the Census; he is very much aware of all these problems and difficulties, but he does not have the money to collect the kind of information needed for an understanding of the geography of second homes in the United States.

[30] George K. Lewis, "Population Change in Northern New England," *Annals,* Association of American Geographers, Vol. 62 (1972), 307–22. The phenomenon is not unique to North America; on June 16, 1973, the *New York Times* quoted a Swiss official as saying that "mass tourism and the second-home movement are threatening to create a huge Alpine slum from Munich to Milan."

[31] Forty-two percent of the people questioned in a statewide poll in 1972 favored closing Vermont to all immigrants from other states; *Newsweek,* July 10, 1972, p. 86.

attracted get-rich-quick speculators as well as legitimate land developers. Reputable developers have built dams to create man-made lakes near major urban centers, and then subdivided the land around them for home sites.[32] Unfortunately, hordes of unscrupulous promoters have used intensive advertising and high-pressure sales tactics to exploit this new and naive market, and they have perpetrated a welter of abuses on unsophisticated buyers.[33] At least 25 speculative, large-scale recreational subdivisions of more than 1,000 lots each have been created in northern California alone, and 40,000 to 50,000 new lots in mountain subdivisions have been created each year, although less than 3 percent have actually been built on.[34] Miles of unneeded and unused access roads have been gashed across the land, destroying vegetation, disturbing wildlife habitats, accelerating erosion, muddying streams, and irretrievably scarring the landscape.

FROM BARREL STAVES TO BOGNORS.   Skiing, the most popular outdoor winter sport in the United States, did not really catch on in this country until 1933, when an ingenious Yankee mechanic in Woodstock, Vermont, hitched a stout rope to an old Model-T Ford engine and invented the first rope tow to haul enthusiasts back uphill after they had skidded their way to the bottom. The railroads started running special weekend ski trains from eastern cities to nearby ski areas, but they were halted by the war, and the big ski boom did not come until after World War II. By 1970 ski areas were widely scattered over the north and west, but the major concentrations were in southern New England and in the Mont Tremblant and Catskill areas (Fig. 11–8).[35] A useful distinction can be made between weekend ski areas, which have limited facilities and draw their clientele primarily from nearby cities, and major ski resorts, which have better conditions and facilities for skiing, but must draw their customers from much greater distances.

A successful weekend ski area must have a chalet or central lodge with a bar and restaurant, a shop where equipment can be bought or rented, a snow-making machine, one or more lifts, a suitable parking area, and good access roads; it helps to have lights so the slopes can be used

[32] The development and economic impacts of such a lake-oriented second-home community in northwestern Pennsylvania are discussed in Richard N. Brown, Jr., *Economic Impact of Second-Home Communities: A Case Study of Lake Latonka, Pa.*, ERS-452 (Washington, D.C.: Economic Research Service, U.S. Department of Agriculture, 1970).

[33] Anyone who has even the slightest urge to "invest" in one of the new second-home developments should read the funny and frightening story on the new land boom which appeared in the *New York Times* on September 4, 1972. It describes high-pressure sales tactics in harrowing detail, and mentions one company which sold 300,000 land contracts without bothering to tell the purchasers that "some of the land was under water, inaccessible, had not been surveyed, would have no amenities, had no police or fire protection or schools within 60 miles, and had no provision for road maintenance, water, or sewage."

[34] James J. Parsons, "Slicing Up the Open Space: Subdivisions without Homes in Northern California," *Erdkunde*, Vol. 26 (1972), 1–8.

[35] Metropolitan newspapers in the ski belt regularly publish listings of facilities and conditions at ski areas, but the definition of a ski area seems to be rather flexible; John Fraser Hart, "The Three R's of Rural Northeastern United States," *The Canadian Geographer*, Vol. 7 (1963), 13–22.

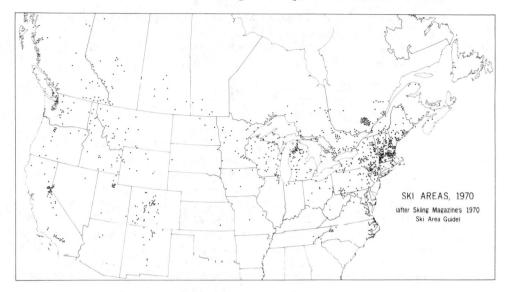

FIG. 11–8.   *Ski areas in the United States in 1970. Compiled from the list in Skiing Magazine's 1970 Ski Area Guide.*

at night. The machines for making artificial snow which have been developed since World War II have been an enormous boon to weekend ski areas, and have permitted their extension southward as far as the Great Smoky Mountains of western North Carolina and eastern Tennessee. The nozzles of the machine spray a mist of fine water droplets across the slope, and these turn to snow before they settle to the ground if the air is cold enough. The machines will work at 32° F., but the ideal is a temperature of 28° F. or lower and low humidity.

Major ski resorts are in areas where longer slopes and heavier snows provide a variety of skiing conditions, but they must also have luxurious accommodations and lively entertainment to attract and hold their visitors.[36] They need easy access by car and plane, large modern lodges, fast and efficient lifts, a selection of bars and restaurants, plenty of shops and boutiques, and good medical facilities.[37] It helps to have facilities for other winter sports, such as skating rinks and bobsled runs, and most ski resorts have attempted to extend their seasons by developing golf courses, swimming pools, tennis courts, riding trails, artificial lakes, and conven-

---

[36] One of my good friends who knows all about such things tells me that the ten leading ski complexes in the United States are Squaw Valley and Heavenly Valley in California, Sun Valley in Idaho, Aspen and Vail in Colorado, Boyne in Michigan, Mont Tremblant in Canada, and Mount Snow, Stowe, and Killington in Vermont. He also told me that skiing is the greatest device for social mixing since the office party.

[37] From a strictly medical standpoint, skiing makes no sense. The injury rate per thousand skier-days is 12 for beginners, drops to five after a week of instruction, but remains three even for experts. In addition to the risks they face on slopes and lifts, skiers are exposed to slips in icy parking lots and falls from ski lodge bar stools; *Time,* February 8, 1971.

tion facilities. They have generated local real estate and building booms for condominium apartments and vacation homes, and they face the same problems as other contemporary urban areas, including traffic jams, air pollution, and the need for planning, building codes, and sewage plants.[38]

THE SNARLING SNOWMOBILE.   The snowmobile erupted across the winter countryside of the North American northlands in the 1960s, and completely revolutionized wintertime outdoor recreation. In 1959 fewer than 2,000 of the low slung, highly maneuverable little vehicles were in operation, but by 1969 around half a million were in use, and the number is increasing each year. The machine is aimed by handlebars connected to a pair of steel skis in front, and propelled at speeds up to 60 miles an hour by a rear-mounted cleated rubber belt driven by a two-cycle air-cooled engine which makes a racket like a demented chain saw.

The snowmobile has been a godsend for those who must live with winter in the northlands, for it enables them to get to places which heretofore they could reach only on snowshoes or skis, and then with difficulty. Utility workers can service lines even when the roads have been blocked by drifting snow, and foresters can continue to cruise timber through the winter. The rancher can haul feed to his starving cattle, ride his fences, and make repairs. No longer need he fear being snowbound, for in an emergency his snowmobile can carry him out, or bring in the doctor he needs. The trapper can patrol a longer trapline, and even the Eskimo has traded in his dog team for a snowmobile. As a crowning blow, in 1969 the Royal Canadian Mounted Police replaced the last of their fabled sled dogs with the new machines.[39]

The snowmobile has also opened up new vistas for family outdoor recreation in winter. At first snowmobiles were used to buzz over the snow-covered slopes of golf courses, tow sleds and skiers, and haul fishing shanties out onto frozen lakes.[40] Snowmobile clubs were formed, expeditions were organized, and some hardier souls even experimented with winter camping. Many resort and summer cottage owners began to keep their places open well into the winter, and the managers of state and national forests were encouraged to develop trails exclusively for snowmobilers. On the other hand, the little vehicles have a perverse way of tipping over, getting stuck, breaking down, crashing through thin ice, leaving litter, cutting fences, mowing down young trees, frightening animals, and rousing the countryside with their infernal racket. They have not been universally well received.

[38] The destructive impact of ski areas on the landscape is dramatically illustrated in *Planning Considerations for Winter Sports Resort Development* (Denver: Regional Forester, U.S. Forest Service, n.d.).

[39] *Newsweek*, March 31, 1969, p. 44.

[40] "As soon as the ice is thick enough (and sometimes even a bit sooner, to their sorrow) the ice fishermen move out and set up their 'cities' on the larger lakes. Many of their 'shacks' are equipped with all the comforts of home—heaters, carpets, easy chairs, and liquid refreshments. Several years ago four enterprising damsels from St. Paul, Minn., made the front pages when the police discovered that they had set themselves up for business in a 'shack' up on one of the larger lakes nearby and were plying quite a profitable trade;" John Fraser Hart and Russell B. Adams, "Twin Cities," *Focus*, Vol. 20, No. 6 (February 1970) 8–9.

The snowmobile is only the most numerous member of a whole tribe of Off-Road Recreation Vehicles (ORRVs) whose use has been hotly controversial: other members include motorcycles, trail bikes, and minibikes; sand buggies and dune buggies; all-terrain vehicles; swamp buggies; and hovercraft.[41] There is no question that some users create excessive noise and leave litter, and that their unwise use can injure and destroy vegetation, harass wildlife and farm animals, and accelerate erosion of trails and stream banks; after all, they are designed to cross terrain that would drive a pack-mule into retirement. Much of the evidence against them, however, is limited, anecdotal, or emotional, and some of the objection to their use may have nasty undertones of class. Those who wish to keep the countryside pure and undefiled often are also those who believe that a bit of suffering is good for the soul, and their joy in attaining an inaccessible spot after a long hard hike can be turned to resentment by the discovery that someone else has already been there on wheels. The use of ORRVs quite clearly is incompatible with certain other uses of the countryside, but it is equally clear that ORRVs have a legitimate claim to the use of some part of it, and this claim must be honored by the establishment of reservations, and the development of trails and facilities for their use.

[41] *ORRV: Off-Road Recreation Vehicles*, A Department of the Interior Task Force Study (Washington, D.C.: Department of the Interior, 1971), p. 13.

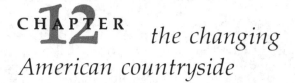

CHAPTER **12** *the changing*
American countryside

Anyone who revisits a familiar rural area in the United States is impressed anew by a paradox: the countryside is becoming emptier, and the countryside is filling up. It is becoming emptier because many old familiar features of the farming landscape are disappearing as technological change makes them obsolete, or farm consolidation makes them surplus, or land abandonment makes them derelict. It is filling up because hordes of onetime city dwellers are moving to the country. The old distinction between urban and rural, between city slicker and country bumpkin, has completely lost what little meaning it might once have had. It is hard to tell where the city ends and the countryside begins. Many paved highways in rural areas, especially those within commuting distance of major cities, have become so built up with nonfarm homes and businesses that they look almost like city streets. Normal commuting distance is not what it used to be; 50 miles no longer seems an onerous journey to work for the people who live in the new homes beside the highways, because they can cover the distance in an hour or so.

The automobile has divorced the worker's residence from his workbench, and it permits him to live where he pleases. It has also divorced many countrymen from their plows and tractors by allowing them to take jobs in nearby cities. The sometime farmer who lives back from the highway may form a car pool with a few of his neighbors. Each drives his own car to the highway, parks it there for the day, and they all make the journey to work in a single car. During working hours one commonly sees clusters of cars parked at each intersection along many highways. What has happened to the land once farmed by their owners? Only the very best has been consolidated, by rent or by purchase, into full-time farm operations, because agricultural production has become more and more concentrated on our better land. Some of the rest is still farmed in desultory, part-time fashion, but much of the land in rough hilly areas has simply been abandoned and allowed to grow up in weeds, briers, and brush.

The very notion of farmland abandonment is anathema to most modern Americans, who still think of our land in terms of clearance, not abandonment. The hardy pioneer, who felled the forest with his trusty axe, and brought the land into cultivation, is one of our national folk heroes, and we feel that somehow we have let him down when we discover than an area larger than the entire state of Iowa has gone out of farming in the past few decades. Most of the eastern United States will revert to woodland within 50 years or so unless man does something to prevent it, and man has not seen fit to prevent it along the spine of Appalachia, in that broad belt of rugged hills and mountains which runs northeastward from Alabama into northern Maine and beyond. The land is too steep and stony for modern farm machinery, and Man has allowed Nature to foreclose her mortgage upon it. Even Mother Nature, however, can only do so much with land that has been worn and battered by two centuries of human use and abuse, and the forests she is putting back on the land are much inferior to those which originally cloaked it.

American farmers are using less land, and their numbers are steadily declining, but they are producing ever larger amounts of food and fiber. In 1790 only one of every 20 Americans lived in a town or city; in 1970 fewer than one of every 20 lived on a farm. Each man-hour of farm labor produced three times as much food and fiber in 1970 as it had produced in 1950. This explosion of agricultural production has been the result of mechanical, chemical, and biological innovations. Machines have replaced horses, and the old horse barn is rapidly fading from the scene; a modern farmer must have at least two tractors, one to haul with and a second to serve as a mobile power unit. The farmer as chemist uses herbicides to control weeds, pesticides to control insects, and gets much of the productivity of his soil out of a sack of fertilizer; American farmers used 20 million tons of commercial fertilizer in 1950, 40 million tons in 1970. The oft-told tale of hybrid corn is but one of many biological breakthroughs scored by plant and animal breeders; now some grave and solemn scientists have even begun to tinker with the sex habits of insects, and they have gotten the poor little things so confused that they are not able to breed at all.[1]

Innovations in agricultural technology have changed farming from a way of life into a complex business demanding a wide range of managerial skills, including the skill of money management. A modern farmer must be able to handle indebtedness as easily as he handles a tractor,

[1] Hal Higdon, "New Tricks Outwit Our Insect Enemies," *National Geographic Magazine,* Vol. 142 (September 1972), 380–99. Mechanical, chemical, and biological innovations in agriculture often are interdependent. As one dramatic example, rice yields in southeastern Texas were increased from 3,100 to 4,500 pounds per acre in only three years by combining a new variety of rice with a chemical weedkiller which controlled grass in the rice fields. Engineers have been designing better machines to pick tomatoes while plant breeders have been designing new tomato varieties which are easier for the machines to pick.

because the farmer who gets ahead is the one who is not afraid to borrow large amounts of money to finance his farm operations. The old general farm which produced a little bit of everything is as dead as the dodo. Today's successful farmer must concentrate his energies and his resources on intensive production of the few items he can produce best. If he grows crops, he lets some other farmer worry about feeding them to livestock; the man who fattens livestock depends upon a crop man to produce his feed.

Some of the most spectacular examples of agricultural specialization are in the irrigated areas of the West, where the largest dry lots for fattening beef cattle cover an area larger than 500 football fields and contain more cattle than the entire state of Delaware, 40,000 plus at any given moment. Dairy farmers in Delaware and other parts of the Northeast have taken the hint, and most of them have stopped trying to raise the feed they need for their animals; they buy alfalfa pellets from Colorado, sugar beet pulp from Minnesota, citrus pulp from Florida, and any other kind of feed that doesn't cost too much. Dry-lot feeding is on the increase even in the dairy country of Wisconsin, where farmers have begun to harvest their forage crops precisely when they are ripe, and store them as hay or silage until the cattle need them. No pasture is wasted because the animals have trampled or soiled it, and many farmers are pulling down the fences they no longer need.

Specialized agricultural management reaches its extremes on vertically integrated poultry farms and in the citrus groves of Florida. The poultry farmer provides the buildings and labor, but the feed company provides chicks and feed, and maintains complete control of the operation. The farmer receives a guaranteed price, which protects him against the wild fluctuations of the poultry market, but he is not allowed to make any management decisions, and he feels like a paid laborer on his own land. In contrast, the man who owns a citrus grove in Florida need never set foot on his land; he hires a production company to tend the grove, market the fruit, deduct the costs, and send him a check at the end of each year.

Farming today is remarkably similar to manufacturing in the middle of the seventeenth century; it is still a home workshop industry, organized into a large number of small and often not very efficient units. The new technology has increased productivity enormously, but costs have risen right along with production, and profit margins have remained virtually unchanged. Individual units must be enlarged to increase their volume of business, and new organizational concepts are necessary; neither vertical integration, which reduces the poultry farmer to the status of a laborer on his own land, nor the citrus industry, which relegates the owner to the passive role of stockholder and coupon clipper, seems to provide a particularly good model for the future.

One fact is clear: most farms in the United States are already too small, and the minimal size for a successful farm operation in this country is steadily increasing. The man who owns an undersized farm may simply walk off and leave it, or he may choose to become a part-time farmer or a part-owner farmer. The part-time farmer finds himself a job off the farm, works at it eight hours a day, uses his morning, evening, and weekend hours on the farm, hires a neighbor to do some of the chores, and takes his vacation when the workload on the farm is heaviest. This kind of routine begins to get old after a while, and eventually the part-time farmer decides to lease most of his land to a neighbor who is a part-owner farmer.[2]

The part-owner farmer owns only part of the land he farms, and leases the rest. Some landowners are reluctant to sell land which has been in their families for generations, and the part-owner farmer probably could not afford to buy the land even if they were willing to sell it, so a leasing arrangement works out best for both sides. The man who buys or leases land to expand his farm operation has to take what he can get, and one of the results of part-owner farming has been the creation of farm operations consisting of several widely scattered blocks of land. Heinous though the idea may be to traditionalists, the progressive modern farmer is a tenant, and he is in debt; he conducts at least part of his operation on leased land and with borrowed money.

The declining acreage of farmland, the increasing size of farms, and the declining farm population have worked serious economic hardships on the hamlets and villages which grew up to serve the needs of local rural people. Any decline in the population they serve almost inevitably means a decline in their business. Furthermore, the automobile has enabled rural people to bypass them completely, and to shop in larger places which offer greater variety. These small market centers look dead when you drive through them, but in fact their population has not declined, despite their stagnating economics. Their houses still provide cheap homes for long-distance commuters and retired people, and many are supported by at least one surprisingly prosperous business started by a local boy who does not want to leave the old home town.

Most villages and small towns used to dream of attracting industry, and some still do, but that dream has been pretty well exploded by the realization that most factories have quite specific needs which can be satisfied only at select locations. Recreation has replaced industry as the universal panacea, and every wide place in the road, no matter how remote or how dismal, now dreams of the day when it will become a great recreational center.[3] The development of a recreational center, unfortunately, takes lots and lots of money. Most city folk will not be

[2] The Eighteenth Edition of *Hobby Publications*, issued by the Superintendent of Documents in 1966, lists *Part-time Farming* (Farmers Bulletin 2178) with the explanation that "part-time farming has become the hobby of many city folks who like to putter around in the country," although the Bulletin itself makes no reference to part-time farming as a hobby. The chances of establishing a small farm home in the California countryside range from "impossible" to "very difficult"; Edward A. Yeary and Edward J. Johnson, *The Small Farm Home*, Leaflet 210 (Berkeley, Calif.: California Agricultural Experiment Station, 1973).

[3] To cite but one example of many, "this rugged area, whose terrain precludes much industrial development, is thus a logical target for the development of recreation as an industry"; Neil Walp, "The Market for Recreation in the Appalachian Highlands," *Appalachia*, Vol. 4, No. 3 (November-December, 1970), 27.

satisfied with the make-do of rural living; accommodations must be improved, because Americans expect to live better when they are travelling than they do at home. Furthermore, city folk grow bored fairly quickly when left to their own devices in the country, and they demand entertainment; nature alone is not enough for them. Even if all the necessary recreational facilities are developed, however, recreation still may not prove to be the boon anticipated, because it is a highly seasonal activity, and during a brief period of three or four months the local people must milk the tourists of enough to live on for the rest of the year.

The major rural recreational areas in the contemporary United States are campgrounds in our parks and forests, and colonies of summer cottages along the shores of our water bodies; water seems essential for outdoor recreation in summer. In winter the most popular outdoor sport for adults is skiing, whose impact has been great but upon relatively small areas. The snowmobile has affected larger areas, and it certainly has been much more controversial, but in time it will come to be accepted as the wintertime equivalent of the powerboat.

City people who use the countryside can do much to help maintain its beauty and prevent its deterioration by spending their money to keep it attractive, whereas most working farmers can no longer afford the luxury of doing so. Most city folk who own land in the country can do their bit, but by far the most impressive examples are showplace gentleman farms; nearly every major city has a few, but their greatest concentration is in the Gentleman Farm Belt which lies just west of our largest city, Megalopolis. The gentleman farmer likes to gaze out on well-kept fields and pastures and fine livestock. His ability to pay for what he likes has created some of the most beautiful rural landscapes in the United States.

The ugliest "rural" areas in the United States are those with which many city folk are most familiar, the hideously predictable highwayside strips, endless avenues of gaudy buildings, garish signs, blinking lights, flapping pennants, and blatant billboards, monotonously flashy look-alikes from coast to coast, whether filling station or drive-in theater, Holiday Inn or Howard Johnson's, Red Barn or McDonald's, Burger King or Dairy Queen, A & W Root Beer or good old Colonel Sanders Kentucky Fried Chicken. They huddle as close to the highway as possible, screening off the countryside behind them, and those who stick to the main highways, especially main highways between major cites, see little more than the screen. Back of the screen, if they will but have the courage to explore, they will find lots of open and empty country.

Preposterous though the idea may seem to denizens of Long Island or Los Angeles, the cities in this country are not about to swallow up our farmland; in fact, they do not take up much more land than is planted to our single leading crop, corn. In 1967 our total built-up area—cities, towns, villages, airports, highways, and the like—covered about 61 million acres, less than 3 percent of our total land area, and only slightly more than the 60 million acres that were planted in corn. If the growth rate of the preceding decade is maintained until the end of the century, no more than 100 million acres, roughly 4 percent of our total area, will be built-up by the year 2000, leaving 96 percent of the United States for nonurban uses, and for enjoyment by people like you and me.[4]

[4] The situation is no different in densely populated England and Wales, where "urban land amounts to only about 11 percent of the total land surface" in 1960, and "by the year 2000 the total area of urban land will probably not occupy more than 15-16 percent of the whole land surface"; Robin H. Best, "Competition for Land Between Rural and Urban Uses," in *Land Use and Resources: Studies in Applied Geography (A Memorial Volume to Sir Dudley Stamp)*, Special Publication No. 1 (London: Institute of British Geographers, 1968), pp. 89–100; quotations from pp. 91 and 98.

Index